AF803164

Innovative Education Technologies for 21st Century Teaching and Learning

Innovative Education Technologies for 21st Century Teaching and Learning

Edited by
Muhammad Mujtaba Asad, Fahad Sherwani,
Razali Bin Hassan, and Prathamesh Churi

CRC Press
Taylor & Francis Group
Boca Raton London New York

CRC Press is an imprint of the
Taylor & Francis Group, an **informa** business

If MATLAB and SIMULINK appears in the book, then use the following at the top of the disclaimer page. MATLAB® and Simulink® are trademarks of The MathWorks, Inc. and are used with permission. The MathWorks does not warrant the accuracy of the text or exercises in this book. This book's use or discussion of MATLAB® and Simulink® software or related products does not constitute endorsement or sponsorship by The MathWorks of a particular pedagogical approach or particular use of the MATLAB® and Simulink® software.

First edition published 2022
by CRC Press
6000 Broken Sound Parkway NW, Suite 300, Boca Raton, FL 33487-2742

and by CRC Press
2 Park Square, Milton Park, Abingdon, Oxon, OX14 4RN

© 2022 selection and editorial matter, Muhammad Mujtaba Asad, Fahad Sherwani, Razali Bin Hassan, and Prathamesh Churi; individual chapters, the contributors

CRC Press is an imprint of Taylor & Francis Group, LLC

Reasonable efforts have been made to publish reliable data and information, but the author and publisher cannot assume responsibility for the validity of all materials or the consequences of their use. The authors and publishers have attempted to trace the copyright holders of all material reproduced in this publication and apologize to copyright holders if permission to publish in this form has not been obtained. If any copyright material has not been acknowledged please write and let us know so we may rectify in any future reprint.

Except as permitted under U.S. Copyright Law, no part of this book may be reprinted, reproduced, transmitted, or utilized in any form by any electronic, mechanical, or other means, now known or hereafter invented, including photocopying, microfilming, and recording, or in any information storage or retrieval system, without written permission from the publishers.

For permission to photocopy or use material electronically from this work, access www.copyright.com or contact the Copyright Clearance Center, Inc. (CCC), 222 Rosewood Drive, Danvers, MA 01923, 978-750-8400. For works that are not available on CCC please contact mpkbookspermissions@tandf.co.uk

Trademark notice: Product or corporate names may be trademarks or registered trademarks and are used only for identification and explanation without intent to infringe.

Library of Congress Cataloging-in-Publication Data
Names: Asad, Muhammad Mujtaba, editor. | Sherwani, Fahad, editor. | Hassan, Razali Bin, editor. | Churi, Prathamesh, editor.
Title: Innovative education technologies for 21st century teaching and learning / edited by Muhammad Mujtaba Asad, Fahad Sherwani, Razali Bin Hassan, and Prathamesh Churi.
Other titles: Innovative education technologies for twenty-first century teaching and learning
Description: First Edition. | Boca Raton ; London : CRC Press, 2022. | Includes bibliographical references and index.
Identifiers: LCCN 2021021809 (print) | LCCN 2021021810 (ebook) | ISBN 9780367699048 (Hardback) | ISBN 9780367699079 (Paperback) | ISBN 9781003143796 (eBook)
Classification: LCC LB1028.3 .I5255 2022 (print) | LCC LB1028.3 (ebook) | DDC 371.33/468—dc23/eng/20211021
LC record available at https://lccn.loc.gov/2021021809
LC ebook record available at https://lccn.loc.gov/2021021810

ISBN: 978-0-367-69904-8 (hbk)
ISBN: 978-0-367-69907-9 (pbk)
ISBN: 978-1-003-14379-6 (ebk)

DOI: 10.1201/9781003143796

Typeset in Times
by codeMantra

This book is dedicated to Professor Nisar Ahmed Siddiqui (Late) and all the Educators and Researchers in the field of Educational Technologies.

Contents

Preface

In the 21st century, classrooms technology can help to transform the classroom experience from a classic teacher-centric into a student-centric experience, where students take a more active role in their learning. Technology also provides teachers and students with access to a variety of educational resources that inspire creativity, critical thinking, communication, and collaboration.

In this modern era of innovation, where all sectors and domains of lives became advanced due to the integration of digital technologies, to cope with this challenging situation, educators are also in need of upskilling themselves to develop digital competencies that are required for 21st-century teaching and learning process. Therefore, this edited book will highlight all the aspects of innovative 21st-century educational technologies and skills which can enhance the teaching and learning process on a broader spectrum based on the best practices around the globe. Also, the challenges and how those challenges can be mitigated by technology which educational institutions are using due to the COVID-19 pandemic for teaching and learning will be highlight in this book.

The silent features of this book is as follows:

- Highlighting the major innovative technologies as required for Education 4.0.
- Teacher competencies for 21st-century education.
- Addressing international cases and senators towards the integration of Educational Technology.
- Highlighting the challenges faced by educational institutions due to COVID-19.
- Mitigations through educational technologies.
- New framework for asynchronous communication for 21st-century education.
- Innovative tools and approaches for teaching with technology.
- The new layout of techno-pedagogy.

The book contains 13 chapters written by various authors from different countries including India, Pakistan and Malaysia. A short description of each chapter is given below.

Chapter 1 highlights the importance of digital competency in ensuring a sustainable, future-proof technical education in coping with the needs of future workplace. In particular, contributors of the chapter wish to focus on how professional digital competency can be developed through the adoption of open educational resources (OER) and open-source software (OSS). **Chapter 2** is an experimental study which was conducted to train the prospective teachers with the help of Virtual Reality (VR) in a public sector university in Pakistan. The study strived to bring innovation in conventional practices and introduced learning with VR. The findings suggest integrating innovative technologies including virtual reality for both teachers' training and students' learning. The purpose of **Chapter 3** is to investigate learners' satisfaction, needs, and challenges with emergency remote teaching such as connectivity, devices, and learning strategies in a prominent public higher education institution during the COVID-19 pandemic. **Chapter 4** discusses the learning outcomes of graduate students in a virtual chemistry laboratory and live demonstrations performed in the traditional laboratory.

Chapter 5 discusses implications for the expansion of Computer-Assisted Language Learning in Pakistan, focusing on ways to pave the way towards the blending of technology in any way, either asynchronously or synchronously, into the ESL context in Pakistan. **Chapter 6** highlights the role of key technological drivers such as ICT adoption, social IoT, and artificial intelligence in e-learning systems with an explanation of the mediating role of pedagogical digital competence, computer self-efficacy and moderating role of technostress and techno overload during COVID-19 in developing country like Pakistan. **Chapter 7** studies the effects of traditional and digital game-based teaching approaches on the mathematics problem-solving attitude of K-8 students. **Chapter 8** discusses the scope of integrating blended mode of teaching-learning to assess the evidence about the quality of the student learning experience and learning outcomes. An account of the relationships between different modes of learning, i.e., face-to-face, on-line, and blended mode, would be given. More specifically, the concerns that need to be addressed at the teacher, student, and institutional levels for effective implementation of blended teaching-learning mode are also further explored.

Chapter 9 explores the journey of faculties at the Department of Education of Sukkur IBA University in shifting to online mode of teaching with a focus on the challenges they faced and the lessons they learnt in order to make their online teaching more effective for teachers. Using the activity theory framework, the research focuses on the contradictions arising as a result of shifting to online mode and how those contradictions were addressed to improve the activity. **Chapter 10** studies the Electroencephalogram (EEG) Signals for Modern Educational Research in engineering education. **Chapter 11** investigates the interrelationship between student's psychological distress and academic performance in academic settings in Pakistan. Additionally, the study further seeks to explore the mediation role of e-learning readiness and moderation of change-oriented leadership in Higher Education Institutions (HEIs). **Chapter 12** contends that while there surely is a great potential and ability in the digital to serve many large-scale human/societal solutions, its wholesale one-size fits all 'borrowing' as a 'panacea' for educational access and continuity needs to be critically debated and scrutinized, especially in the context of education in the global south. **Chapter 13** proposes a research study which assists in indicating the level of motivation among educators towards the integration of Education 4.0 in their teaching at Community Colleges and Schools in rural Sindh, Pakistan. Furthermore, the outcome of this chapter helps the stakeholders and educational reformers to implement education 4.0 in educational institutions.

This book has an integrated approach of delivering the concept of higher education through two aspects, viz. technologies and innovative tools. We hope that this book will create interest among the researchers who are working in similar fields.

Thank you.

Dr. Muhammad Mujtaba Asad
Head of Educational Technologies and TVET Research Group.
Sukkur IBA University (SIBAU), Airport Road, Sukkur, Pakistan.

Dr. Fahad Sherwani
Assistant Professor
National University of Computer and Emerging Sciences, Karachi, Pakistan

Dr. Razali Bin Hassan
Associate Professor
Universiti Tun Hussein Onn Malaysia, Parit Raja, Batu Pahat, Malaysia

Prof. Prathamesh Churi
Assistant Professor,
School of Technology Management and Engineering,
NMIMS University, India

MATLAB® is a registered trademark of The MathWorks, Inc. For product information, please contact:

The MathWorks, Inc.
3 Apple Hill Drive
Natick, MA 01760-2098 USA
Tel: 508-647-7000
Fax: 508-647-7001
E-mail: info@mathworks.com
Web: www.mathworks.com

Acknowledgements

First and foremost, praises and thanks to God, the Almighty, for His showers of blessings throughout our work to complete the book **Innovative Education Technologies for 21st Century Teaching and Learning** successfully.

The authors have been really motivational, and this is the key in establishing CRC Press, Taylor and Francis, as one of the best houses for publication in the subject of Data Protection and Privacy. We thank them all for considering and trusting our book for publishing their valuable work. We also thank all authors for their kind co-operation extended during the various stages of processing of the book in CRC Press, Taylor and Francis.

For the success of any edited book, reviewers are an essential part and hence they merit sincere appreciation. The inputs of reviewers are used in improving the quality of submitted book chapters. The reviewing of a book chapter is essential to assure the quality of the chapter published in any book. We thank the following reviewers for their excellent contributions during the review process:

- Dr. Zaheer Abbas, Universiti Tun Hussein Onn Malaysia
- Ms. Aparna Sharma, NMIMS University, Mumbai, India
- Dr. N. Asokan, Educational Leader, Consultant, Certified Accreditor, India
- Dr. Gul Afshan, Sukkur IBA University (SIBAU), Airport Road, Sukkur, Pakistan
- Dr. Ali Nawab, Sukkur IBA University (SIBAU), Airport Road, Sukkur, Pakistan
- Dr. AlKarim Datoo, Sukkur IBA University (SIBAU), Airport Road, Sukkur, Pakistan

The overwhelming responses from the authors across the world have been a real motivation and support in taking forward this book in the area of Data Protection and Privacy in the healthcare field. Last but not least, we would like to thank Ms. Cindy Renee Carelli (Executive Editor – Engineering, CRC Press / Taylor and Francis) and Ms. Erin Harris (Senior Editorial Assistant – Engineering, CRC Press / Taylor and Francis) for their valuable support in the book-editing process.

Editors

Muhammad Mujtaba Asad is an Assistant Professor and Lead Researcher of Educational Technologies and TVET Research at Sukkur IBA University. Dr Mujtaba has national and international working experience of more than 10 years in the industrial and educational sectors in the domain of Educational Technologies, Work & Product-Based Education, and Technical and Vocational Education. He has published more than 63 research articles (ISI & Scopus Indexed) in reputed international journals and conferences worldwide as well as a book related to the observational skills of competent educators under the CS publishing house USA. He has also received five medals—three Gold, one Silver and one Bronze—in international research and development completions with seven best research awards in Malaysia, Korea, Saudi Arabia, Pakistan, and UAE on his commercialized projects.

Moreover, Dr. Mujtaba is also an active member of 7 national and international research grants. Dr. Mujtaba is also an internationally certified educational technologist (Malaysian Board of Technologist) and holds certifications and qualification in industrial safety and health education from NIOSH and OSH Department of Ontario, Canada. He also serves as an Editorial Board Member to 12 reputed and indexed international journals. Dr. Mujtaba holds several international copyrights and has filed one product patent under the title of HAZFO Expert 1.0 for Drilling Operation. He had been awarded an honorary lifetime International Ambassadorship by University Tun Hussein Onn Malaysia based on his academic achievements and community services. Dr. Mujtaba has also been awarded the title of approved PhD and MPhil/MS supervisor by Higher Education Commission of Pakistan.

Fahad Sherwani is an Assistant Professor at National University of Computer and Emerging Sciences, Pakistan. Dr. Fahad has published several research articles and two books related to artificial intelligence and Educational Technologies.

Razali Hassan is Director of Malaysia Research Institute for Vocational Education & Training (MyRIVET). With a PhD in Education from University of Warwick, UK, Master of Science in (TVET) Uni. Putra Malaysia, and Bachelor of Technology with Edu. (Electrical Technology) Malaysia University of Technology, he also has a certificate of teaching from Technical Teachers Training College. He has worked as a teacher, lecturer, consultant, researcher, and motivator in various ASEAN, Europe, Middle East, and African countries. Dr. Razali has published several research articles and books in Educational Technologies and TVET domains.

Prathamesh Churi (SMIEEE, MACM, LMCSI, LMISTE, MIGIP) is a faculty member in Computer Engineering Department in School of Technology Management and Engineering, NMIMS University. India. Currently, he is an Associate Editor of *International Journal of Advances in Intelligent Informatics* and *International Journal of Innovative Teaching and Learning in Higher Education*. He is also the research mentor in Cerebranium. He is actively involved in the peer review process of reputed IEEE and Springer journals such as *IEEE Transactions on Education and Springer Education and Information Technologies*, as well as 17 other journals. He has 60+ research papers in national/international conferences and journals (Scopus and SCI Indexed) and 5 (including 2 Australian patents) and 4 international books (CRC Press/Taylor and Francis publications) in the field of Data Privacy and Education Technology. He has been a keynote speaker, chair, convener in international conferences including flagship conferences such as IEEE TALE 2017–2020, Springer ICACDS, and the like. He recently received "Best Young Researcher" award from GISR Foundation for his research contribution in the field of Data Privacy and Security, and Education Technology. He has received appreciation award for the best faculty from NMIMS University and was also the recipient of the "Best Reviewer Award" in IEEE TALE 2020. He is an active leader, coach, mentor, and volunteer in many nonprofit organizations and is also involved as board of study member in many universities for curriculum development and educational transformations.

Contributors

Mohammad Jafre Zainol Abidin
University Sains Malaysia
Penang, Malaysia

Donnie Adams
Faculty of Education
University of Malaya
Kuala Lumpur, Malaysia

Gul Afshan
Department of Business Administration
Sukkur IBA University
Sukkur, Pakistan

Ahmed Ali
Electrical Engineering Department
Sukkur IBA University
Sukkur, Pakistan

Unaeza Alvi
Department of Education
Sukkur IBA University
Sukkur, Pakistan

Fatima Bashir
Department of Computer Science
Bahria University
Karachi, Pakistan

M. Bilal
Department of Computer Science
Bahria University
Karachi, Pakistan

Rameen Butt
Department of Education Sukkur Institute of Business Administration
Sukkur IBA University
Sukkur, Pakistan

Muhammad Ayub Buzdar
Government College University
Faisalabad, Pakistan

Bhawani Shankar Chowdhry
Department of Electronic Engineering
Mehran University of Engineering & Technology
Jamshoro, Pakistan

Al Karim Datoo
Department of Education
Sukkur IBA University
Sukkur, Pakistan

Nadira Dayo
Department of Education
The Begum Nusrat Bhutto Women University
Sukkur, Pakistan

Dorothy DeWitt
Faculty of Education
University of Malaya
Kuala Lumpur, Malaysia

Prashant Gupta
Department of Plastic and Polymer Engineering
Maharashtra Institute of Technology
Aurangabad, India

Sumera Hanif
Government College University
Faisalabad, Pakistan

Hamid Ikram
Government College University
Faisalabad, Pakistan

Payal B. Joshi
SVKM's NMIMS University
Mumbai, India

Maryam Kalhoro
Department of Business Administration
University of Sindh
Jamshoro, Pakistan

Waqas Khan
Faculty of Business & Management Sciences
The Superior University
Lahore, Pakistan

Trishul Kulkarni
Department of Mechanical Engineering
Maharashtra Institute of Technology
Aurangabad, India

Fakhr-Un-Nisa Kurio
Department of Education
Sukkur IBA University
Sukkur, Pakistan

C.S. Lai
Faculty of Technical & Vocational Education
Universiti Tun Hussein Onn Malaysia
Parit Raja, Malaysia

M.F. Lee
Faculty of Technical & Vocational Education
Universiti Tun Hussein Onn Malaysia
Parit Raja, Malaysia

S.C.J. Lim
Faculty of Technical & Vocational Education
Universiti Tun Hussein Onn Malaysia
Parit Raja, Malaysia

M. Marouf
Department of Computer Science
Bahria University
Karachi, Pakistan

Quratulain Mirza
Mehran University of Engineering and Technology
Jamshoro, Pakistan

A. Nawab
Department of Education
Sukkur IBA University
Sukkur, Pakistan

Qasim Ali Nisar
Faculty of Business & Management Sciences
The Superior University
Lahore, Pakistan

T. Noor
Department of Education
Sukkur IBA University
Sukkur, Pakistan

Habibullah Pathan
Mehran University of Engineering and Technology
Jamshoro, Pakistan

Sidra Shehzadi
School of Multimedia Technology and Communication
Universiti Utara Malaysia
Changlun, Malaysia

Hammad Siddiqui
Department of Education Sukkur Institute of Business Administration
Sukkur IBA University
Sukkur, Pakistan

Saira Sohail
Faculty of Business & Management Sciences
The Superior University
Lahore, Pakistan

Raheel Ahmed Soomro
Department of Education Sukkur Institute of Business Administration
Sukkur IBA University
Sukkur, Pakistan

Toufique Akbar Soomro
Department of Electronic Engineering
Quaid-e-Awam University of Engineering, Science and Technology
Larkana, Pakistan

Bambang Sumintono
Faculty of Education
University of Malaya
Kuala Lumpur, Malaysia

Sahib Khatoon Thaheem
Mehran University of Engineering and Technology
Jamshoro, Pakistan
and
University Sains Malaysia
George Town, Malaysia

Bhagwan Toksha
Department of Basic Sciences and Humanities
Maharashtra Institute of Technology
Aurangabad, India

Nasima Zainulabidin
Department of Education
Sukkur IBA University
Sukkur, Pakistan

1 Toward Future-Proof Technical Education

Digital Competency Development Through Open Educational Resources & Software

S.C.J. Lim, M.F. Lee, and C.S. Lai
Universiti Tun Hussein Onn Malaysia

CONTENTS

1.1 INTRODUCTION

With the advent of industrial revolution (IR) 4.0, the process of digitalization is happening rapidly in all sectors of industry and business. Across many industries, companies are rapidly transforming their business operations by adopting information and communication technology (ICT) tools, which are used for the collection, processing, and analysis of their customer and business operation data digitally. The adoption of digital tools in business-related data analysis allows for the precise discovery of customer preferences that, when applied correctly, may help to boost the next sales growth. Manufacturers are progressing toward automating their shop floor production and are adopting ICT technologies such as Enterprise Resource Planning (ERP) to make the entire process manageable and smarter with real-time feedback. Such a digital transformation helps to reduce production downtime and

DOI: 10.1201/9781003143796-1

improve product time-to-market in a cost-effective manner. Digitalization strategies help companies to gain valuable insights from their existing business processes for sustaining the competitive advantage in the market. ICT technologies serve as an enabler to empower the next wave of business growth as it allows efficient delivery of products and services in a timely and cost-effective manner.

This rapid digital transformation in the industry has resulted in the rising demand for a workforce that is not only technically competent but also possesses the necessary digital skills to keep up with the industry's ever-changing needs. Recently, Asia-Pacific Economic Cooperation (APEC) (2020) has stated that digital skills are in high demand across occupations, industries, and economies, particularly specific digital skills such as customer relationship management (CRM), digital design and marketing, manufacturing, data analysis, computer and networking support, software, and programming. In the context of Malaysia, Global Skills Index 2020 (Coursera 2020) reports that Malaysia is at the competitive level for the business trending skills like digital marketing but emerging level for technical skills such as those related to data science, cloud applications, and artificial intelligence (AI). Malaysia earned a 4.5 out of 10 on the Hays Global Skills Index, an annual evaluation of the problems and developments affecting the professional labor market, and the key factor that contributes to this figure is that fewer youth are able to join the labor market due to the demand on digital capability (Gho 2019). In addition, JobStreet.com Malaysia (2019) also reported in its Job Outlook 2019 survey that employers are focused on hiring talent with digital skills. With possible replacement of conventional job functions in the future toward automation, digitalization and AI-based, all these are strong indications that digital competency is an important prerequisite for securing future employments.

From the educational perspective, digital transformation is also required to equip students with the necessary digital skills and computational thinking to become the next generation of digital innovators. The situation is further accelerated due to the recent global pandemic of coronavirus disease (COVID-19), where digital presence is prominent, and with teaching and learning activities dependent almost entirely on digital platforms and tools. The dramatic changes in the daily life due to the pandemic have highlighted the importance of digital competency as a necessary skill of survival. While higher education institutions realize the need to undergo digital transformation to ensure graduates future employability, most of them are struggling to keep up with the latest development. The shifting focus toward digital teaching and learning has resulted in institutional decision-making for issues such as digital content creation, digital competency training for both educators and students, career development, and also information and communication technology (ICT) facilities upgrade, which may require substantial financial budget.

To ensure a sustainable and future-proof technical education, this chapter wishes to discuss how open educational resources and software will aid the digital competency development in coping with the needs of future workplace. In particular, the chapter shall review the recent trends and development in digital competency, open educational resources (OER), and open-source software (OSS). Based on the status quo and suggestions on how to bridge the gap among digital competency, OER and OSS shall be presented, along with brief recommendations for future works.

1.2 DIGITAL COMPETENCY

The rapid development in the digital world has dramatically changed our daily lifestyle in multiple aspects. These changes make digital competency a crucial predictor of an individual's progress, whether in schooling or in career growth. Thus, it is important to have a common view of the characteristics, domains, or concepts of digital competency so that all the parties involved will take their responsibility and take part in country development in the modern digital era.

According to World Economy Forum (2016), the gap between what people learn in school and what they need to succeed is widening as formal education falls far short of providing students with the skills they need to succeed. They address three domains with 16 crucial proficiencies that today's students need to succeed in their careers in the 21st century. The three domains are (1) foundational literacies (*Literacy, Numeracy, ICT Literacy, Scientific Literacy, Financial Literacy, Cultural and Civic Literacy*); (2) competencies (*Critical Thinking/Problem-Solving, Creativity, Communication, Collaboration*); and (3) character qualities (*Curiosity, Initiative, Persistence/Grit, Adaptability, Leadership, Social and Cultural Awareness*). From our extensive literature survey, there are various digital competency frameworks or standards contributed by various countries and scholars. Tables 1.1 shows a summary of available digital competency framework and standards, sorted according to years of introduction.

In general, digital competence emphasized the integration of other fields and the development of the term in professional settings, as opposed to digital literacy, which encompasses a wide range of viewpoints, ranging from technical "know-how" to cognitive skills to social activities and constructive participation with digital knowledge (Spante et al. 2018). From Table 1.1, it is observed that digital competency does not only emphasize on the technological aspect but also consists of others element to be in line with the demand of the career needs of the 21st century. Most of the frameworks or standards shown in Table 1.1 are more inclined to guideline or references for mastering the essential skills required, particularly in the 21st century.

Table 1.2 provides an in-depth comparison among the reviewed digital competency frameworks and standards. In contrast, the Malaysian DCS is slightly different as it is an evaluation rubric to ensure the Malaysian government aim being achieved, which is to prepare the talent funnel to feed the needs of digital economy jobs (Zainudin 2016). Ministry of Education Malaysia has developed a Digital Competency Standards (DCS) as an assessment tool to determine primary and secondary school students' digital skills; this was started in 2016 and focused on three domains: Technology, Digital Citizenship, and Cognitive.

On the other hand, the DQ Framework developed by DQ Institute Singapore contains the most comprehensive coverage. This framework provides a common understanding, structure, and taxonomy as a point of reference by integrating over 25 leading digital competency frameworks worldwide with metrics (Park 2019). DQ Framework claims to be the first global standard for digital literacy, digital skills, and digital preparation that is approved by IEEE Standards Board as the IEEE 3527.1TM Digital Intelligence Standard (DQ) in the year 2020 (Park 2019). The framework can be used as a skills-map that shows the relationship between skills and sectors in the global arena.

TABLE 1.1
Summary of Digital Competency Framework and Standards

Framework /Standard	Description
Malaysia DCS standard (Zainudin 2016)	• A tool developed by Ministry of Education, Malaysia, to assess primary and secondary schools students' digital skills.
ISTE Standards for Educators (Crompton 2017)	• A standard suggested by International Society for Technology in Education (ISTE) that focuses on general indicators specifying teacher digital skills with essential condition. • This is a comprehensive framework that encourages all the important parties in education to take part in order to create an effective learning environment through technology.
INTEF CDCFT Framework (National Institute of Educational Technologies and Teacher Training 2017)	• A framework by the Sepanyol National Institute of Educational Technologies and Teacher Training (INTEF). • This framework promotes continuous professional development among teachers by improving their teaching practice that in line with 21st-century needs.
EFT DigCompEdu Framework (Brolpito 2018)	• A recommendation by European Training Foundation (EFT). • This framework provides a general reference for all levels of education that focus on teachers' and trainers' professional development from the aspect of educators' duty.
UNESCO ICT CFT Framework version 3 (Hine 2011)	• ICT Competency Framework for Teachers of UNESCO focuses on providing guidelines for teachers to integrate ICT in class to achieve the Sustainable Development Goals, a vital element in the 21st century. • This framework adds in the current advances in technologies, including the open educational resources (OER) and the inclusive education.
UK ETF DTPF Framework (Education and Training Foundation 2018)	• Digital Teaching Professional Framework (DTPF) is suggested by The Education and Training Foundation of United Kingdom. • This framework aims to support teachers' teaching practice with a comprehensive pedagogy guide that integrate with the suitable technology used.
Canada DCF framework (Ministry of Education and Higher Education Quebec 2019)	• Digital Competency Framework (DCT) is developed by Ministry of Education and Higher Education, Canada. • This general framework not only focus on teachers but also all levels of employees to master the labor market needs for the 21st century.
Singapore DQ framework (Park 2019)	• Digital Intelligence (DQ) Framework is launched by DQ Institute Singapore that highlights human intelligence. • This framework is universal by providing the definition of digital competencies in a global vein that is composed knowledge, skills, attitudes, and values. • It is a general framework that focuses on education and industry.

TABLE 1.2
Comparison of Digital Competency Framework and Standard

Framework/Standard	Characteristics	Focus	Use
DCS (Zainudin 2016)	3 domains with 9 indicators	Student	Rubric
ISTE (Crompton 2017)	7 domains with 3 levels	General	Guideline/reference
CDCFT (National Institute of Educational Technologies and Teacher Training 2017)	5 areas and 21 competences defined in 6 proficiency levels	Teachers	Guideline/reference
DigCompEdu (Brolpito 2018)	6 areas with 23 competences	Teachers and trainers	Guideline/reference
ICT CFT (Hine 2011)	6 aspects with 3 levels	Teachers	Guideline/reference
DTPF (Education and Training Foundation 2018)	7 elements with 3 stages	Teachers	Guideline/reference
DCF (Ministry of Education and Higher Education Quebec 2019)	12 key dimensions	General	Guideline/reference
DQ (Park 2019)	3 levels, 8 areas, and 24 competencies.	Education and Industry	IEEE Standard

1.3 OPEN EDUCATIONAL RESOURCES & OPEN SOURCE SOFTWARE

In alignment with the noble goal of providing education that is both versatile and affordable to cater for the needs of students with different socioeconomic background, the adoption of Open Education Resources (OER) and Open Source Software (OSS) is of crucial importance. On top of that, the advancement of Internet technology has opened up opportunities for online learning which, in turn, fosters enabling environment for distance education and at the same time supplements the conventional face-to-face learning environment. Within the online and distance learning settings, the use of OER and OSS has become an important part of learning process and has become a new trend in the education sector as it provides free access of learning materials to both students and teachers.

In general, OER is defined by UNESCO (2012) as: "teaching, learning and research materials in any medium – digital or otherwise – that reside in the public domain or have been released under an open license that permits no-cost access, use, adaptation and redistribution by others with no or limited restrictions." Likewise, Hewlett Foundation defines OER as the teaching, learning, and research resources that reside in the public domain or have been released under an intellectual property license that permits their free use and repurposing by others (Bliss and Smith 2017). Within the context of OER, users are allowed to use, duplicate, and manage the materials; to distribute and share the contents with others; and to modify and combine with other open educational contents. These advantages of OER have brought made an incredible contribution to the education development across the globe. Some of the significant examples of OERs are Open Courseware (OCW), MOOCs (Massive Open Online Courses), and OER portal.

On the other hand, OSS commonly refers to software where the source-code is available and freely distributed. Users are licensed to use, share, and modify the source-code of OSS and redistribute the modified version of the source-code (Wei 2011), and this is known as Free OSS (FOSS).[1] Nevertheless, it is important to note that the OSS is not necessarily free of charge. The term "free" simply means that the users have the freedom to use, modify, and distribute the OSS. In general, although many OSS are free to use without any cost, some OSS does require minimum pricing. Using OER/OSS for educational purposes may help to reduce the expenditure on software and reference materials for teachers and students. This is due to most of the OER and OSS being reliable, stable, flexible, and, more importantly, free of charge (Lakhan and Jhunjhunwala 2008; Khan and UrRehman 2012). The dimensions of OER/OSS can be viewed from several perspectives as follows:

i. *Quality/Originality*: The main purpose of OER is to support academicians in searching referral materials for teaching and learning and research (McAndrew et al. 2012). As a result, the accuracy and originality of OER are critical in determining the quality of teaching and the reliability of research since those materials are used as a primary source of reference. It is suggested that users should play a more proactive role in OER quality control by evaluating and rating the resources they have used (Atenas and Havemann 2013). By doing so, the sense of inclusive participation in quality control can be cultivated among the users.
ii. *Availability*: OER/OSS should be broadly available to and easily accessible by the potential users. Otherwise, the aims of OER/OSS will be meaningless. To date, there are several modalities of OER available worldwide, such as OCW, OER websites, MOOCs, and repositories (Navarrete, Luján-Mora, and Peñafiel 2016). Both teachers and students can adopt and adapt the relevant contents to support the teaching and learning process either in face-to-face or virtual learning environment. Additionally, there are also many legal and trustable websites available that allows users to download OSS for free.
iii. *Copyright/Licensing*: OER/OSS usually comes together with open license (e.g., General Public License) that grants permission to users to use and share the contents for free without and with minimal restrictions. In many contexts, licensing is no longer an issue because a wide range of creative common licenses related to OER/OSS are already in place (McAndrew et al. 2012). This will help to achieve the aims of sharing, distribution, and reuse of open contents.
iv. *Training*: Since OSS is freely available for public, in most cases, the OSS providers or developers do not provide training on OSS application. In addition, some OSS are not that user-friendly as compared to commercial

[1] Free and Open Source Software (FOSS) often refers to software that are both free (in terms of cost, as in "free" pizza) and free (as in "freedom"). Open Source Software (OSS) does not necessarily refer to free (as in "freedom") software (i.e. source code availability), which is contrary to free-of-charge distribution and usage (i.e., zero cost).

versions because the developers do not pay sufficient attention to the OSS interface design. This can be an issue for users who are not tech-savvy (McAndrew et al. 2012).

1.4 DISCUSSION

We shall discuss the status quo, challenges, and suggestions of how digital competency can be developed sustainably through OSS and OER from a few perspectives as elaborated in the following subsections.

1.4.1 Industry

The ICT industry has been changing rapidly. Contrary to how it was decades ago, where industrial ICT applications were still very much dependent on standalone desktop computers with retail boxed software products, the scenario looks very different nowadays with the introduction of relatively low-cost ICT devices (e.g., laptop computers, smart phones, and tablets) and cloud-based applications (or Apps). All these are made possible due to several factors, mainly ICT technology breakthrough, which includes high-speed mobile networking technology (e.g., from 2G mobile network to the latest 5G mobile network, at the time of this writing), high-speed Internet penetration and user growth (due to heavy investment in ICT infrastructures, e.g. broadband services at a national level and undersea cables at a global level), and worldwide government-focused initiatives that view ICT as basic infrastructure or necessity for the people. As a result, all these have spurred the growth of information technology (IT)-related services, with the emergence of a great number of online, cloud-based applications (in various governmental, industrial, commercial, and educational settings) that has fundamentally changed our modern daily lives.

In relation, since the inception the of smartphone era in 2007 (where the first iPhone was introduced), IT has been emphasized on web- and mobile-based applications (hereafter refer to Apps). Due to the rapid increase in Internet users and affordable device ownership, the Apps market has experienced tremendous growth over the years. In order to ensure high-quality of services and to stay competitive in the market, IT companies have invested a significant amount of resources in capacity building—including installation of large and efficient data centers worldwide, research into ensuring agile software development, efficient deployment, etc. The cornerstone of all these is the adoption of OSS. Besides cost consideration, OSS allows IT companies to customize their own version of software to suit their needs. This allow sustainable growth for IT companies in the future. Major IT companies have been utilizing OSS as a viable solution to power their servers, services, and also software development.

For instance, Microsoft is one of the IT corporations that has embraced open source in their company products. The company believes that open source strategy offers significant advantage to sustain their business and claims that it has become an accepted model for software collaboration across multiple companies (Sawers 2021). Tapping on the large open source communities that are globally distributed, Microsoft is able to innovate their product offerings, especially their cloud-based services such as Microsoft Office 365. In addition, Microsoft acquired GitHub in 2018, a major open-source coding repository/code sharing database and development tool for developers (Bass and Newcomer 2018). This is part of their open source strategy to encourage collaboration among software developers. Another good example is Google, where OSS application and deployment has been emphasized since its inception. Besides adopting Linux-based servers over other expensive alternatives, a few of their most famous product—Android mobile operating system, Chrome browser and ChromeOS—are all based on OSS. Google has also open-sourced its deep learning software library named Tensorflow,[2] a platform for machine learning, to accelerate the coding, development, and adoption of AI in the industries. For software development, both Microsoft and Google have released their OSS-based tools that are free for download and use, such as the Visual Studio Code[3] by Microsoft and Kotlin[4] programming language by Google that focused on Android mobile app development.

For any successful implementation involving OSS, the availability of good user guide, tutorials, and training are essential. Besides the aforementioned software services and tools, IT companies are also offering OER in the form of free tutorial courses and even certification for their services. Microsoft, for example, is encouraging their users to obtain certification in their OER-based learning portal called Microsoft Learn.[5] Google also partnered with Coursera[6] to offer technical certification courses on their products and services.

In a nutshell, the IT industry is actively adopting open-source technology as a viable business solution for their technical needs and also promoting OER as a way to educate the public to become users of their products and services. Since ICT technology has been widely applied and embedded in many areas, we reckon that OSS and OER are indispensable toward achieving digital competency of future workforces. Educational institutions need to embrace these changes and start to look at practical ways to adopt OSS and OER in their curricula. We shall explain the details in the following subsections.

1.4.2 Educational Institution (EI)

Previously, it was a great challenge to create a friendly environment to support the utilization of OSS at the institutional level. This is because utilization of OSS in the competency training context does not solely depend on knowledge and skills of educators, it also requires technical support, peer support, as well as managerial support (Hepburn and Buley 2006). Proper technical support is usually required to ensure the proper implementation of software as not all educators are well-trained to deal with the technical issues and troubleshooting. Yet it is noteworthy that even some technical support teams have inadequate experience and knowledge in dealing with open source technology because they are used to certain commercial/proprietary software in which they have undergone proper, structured training programs to learn

[2] TensorFlow: https://www.tensorflow.org/
[3] Visual Studio Code: https://code.visualstudio.com/
[4] Kotlin Programming Language: https://developer.android.com/kotlin/
[5] Microsoft Learn: https://learn.microsoft.com/
[6] Google Certification Courses on Coursera: https://www.coursera.org/google

the know-how. With the high cost of ownership for both the hardware and software packages, IT education was often available only for the privileged few.

However, the aforementioned situation has changed. As explained in earlier section, ICT devices are getting affordable for the masses. From EI's perspective, this has helped to promote ICT-based education as EIs are able to setup their ICT facilities at a lower cost. Students are also able to own their personal ICT devices. Nowadays, the Bring Your Own Devices (BYODs) for teaching and learning across EI worldwide is already a common scene. Most people own one or more smartphones.

In relation, ownership of computing devices used to be expensive and unaffordable, especially for those in rural or disadvantaged communities. Nowadays, with the advent of low-cost but powerful single-board computers (SBC) like Raspberry Pi[7] and accompanying open source software implementation, ICT education where free resources are freely available online has become commonplace. There are other SBCs available in the market, such as BeagleBoard[8] and Orange Pi,[9] that are equally powerful and affordable. Commonly supported by OSS, the operating system for these SBCs are usually Linux-based or Android-based, which offers great flexibility for users. Raspbian,[10] the Linux-based operating system customized for Raspberry Pi boards, comes free with software packages such as Libre Office, Python programming language, and a suite of educational software such as Mathematica for scientific computing. Besides, there are a lot of OER available online to teach users on multiple aspects of the SBC usage, from software installation to building interesting do-it-yourself (DIY) projects. The market availability of such a board also has allowed students of all levels to better understand and apply computing technology for their own creativity.

All these developments have presented great opportunities for EI to innovate their teaching processes. With cost-effective ICT hardware setup, the adoption of OSS and OER shall further enable EI to sustain in the digital era. In line with the changing industry landscape that emphasizes digital competency, EI can work together with IT companies to offer certification courses for students by utilizing their OSS and OER. Most of these companies offer low-cost or even free educational licenses for students' use. Besides those offered by IT companies, EI also can adopt nonprofit OSS and OER that are available for educational purposes. For instance, the aik12-MIT[11] initiative by MIT media lab, which is an informative learning portal about artificial intelligence (AI) for K-12 students (MIT Media Lab 2020), can be a great reference to learn AI. EIs may also collaborate with existing MOOC platforms (such as Coursera[12]) to create niche certification courses that are available for public. These will help to publicize the EI internationally, improve their enrolment numbers, and also serve as new possible source of income for the institution. With this in mind, EIs should invest more in their human resource (i.e., educators) to enable their best utilization of OSS and OER. We explain the role of educators in detail in next subsection.

---

[7] Raspberry Pi: https://opensource.com/resources/raspberry-pi
[8] BeagleBoard: https://beagleboard.org/
[9] Orange Pi: http://www.orangepi.org/
[10] Raspbian OS: https://www.raspbian.org/
[11] AI in K-12 by MIT: https://aieducation.mit.edu
[12] Coursera: https://www.coursera.org

1.4.3 Educators

Educators play a crucial role in nurturing future digital workers. While OER with specific content information and knowledge is widely available for students, educators are the frontliners in complementing their students' digital learning journey. Decades ago, educators had limited choices in terms of ICT hardware and software. The choice of educators on certain software, especially on commercial/proprietary software, very much decides the learning experience of their students. In most cases, teachers stipulate the use of specific software for learning purpose, and students often do not have freedom to have other choices. Mainly, this is due to the scarcity of resources for training the educators to use FOSS (Thankachan and Moore 2017), which lead to motivational problems on software adoption.

Today, with the availability of OSS and associated OER, the situation has greatly changed. As explained in previous sections, IT corporations are adopting the open-source model in their software development, and they value the contribution of their user communities. The adoption of such a model has a few advantages, such as user community building that starts from the EI, ensuring future product subscription and customer loyalty. These corporations also have the resources to create well-planned, interactive learning contents that are associated with their products. On the other hand, the non-commercial FOSS has been a community-focused effort since the beginning. The challenge of FOSS lies in its software maintenance, which is commonly done by a software users community or by an organization (for instance, Mozilla Foundation, a nonprofit organization that maintained the Firefox browser). However, it does not mean that FOSS are of low quality. The key lies in the frequency of software updates (e.g., stability and quality), availability of software tutorial, and comprehensive user documentation, which directly impacts software adoption. With the rise of social media and content creators (e.g., YouTubers), there are also a great number of free tutorials and OER available on FOSS. A good example is the video tutorial on OBS Studio, a FOSS video creation and broadcasting software, that was popular on YouTube during the pandemic. In short, all these indicate that users nowadays have more freedom to choose the software that suits their needs.

With the aforementioned background, the role of educators are changing. Regardless of whether the software is commercial or not in nature, the more important aspect is that educators are able to determine what's best in their situation. The important factors of consideration are as follows: availability and quality of OSS and OER, and whether they really suit the needs of their teaching and learning sessions. Educators need to have the competency themselves to differentiate among the available options. Their role is no longer focused solely on content delivery and pedagogical skills but more on mentoring students in their learning journey toward the required digital competency. Educators shall embark on the learning journey together with students, guiding and supporting students whenever they require it. This is important as the know-how of digital competency is constantly updated, in line with new or improved functions or features over time.

The availability of OER and associated OSS is powerful as it allows educators to focus on the learning theme instead of the content. For instance, to teach "computational thinking" as the learning theme, an educator shall be able to set suitable learning objectives/milestones, searching and evaluating available OER and OSS, and planning an achievable, goal-based learning path for students. In other words, future educators should be the advocate for adaptive syllabus or personalized curriculum for individual students. Such a learning style carries a few advantages: it is highly personalized to suit an individual student's progress, students are able to locate additional information to aid understanding, their learning progress are monitored, and it is also good motivation for both students and educators due to continuous learning.

Relatedly, although the combination of OSS and OER is useful toward achieving digital competency, it is widely distributed across the Internet. Good resources may be difficult to locate, and there can be steep learning curves to digest the materials before an educator is able to apply it for teaching. Thus, instead of delivery of OSS know-how, educators need to have the competency to discover useful OSS and OER for their own needs. If suitable OER is not available, it is also becoming increasingly necessary that they can play a role as a content creator and share it with the community. It is obvious that these competencies are essential for educators of the future, and EIs should focus on training their educators to acquire such competencies. In this context, the government's educational policies need to keep up and be aligned with the EIs initiatives in the direction toward digital competency.

1.4.4 Government

In general, the Malaysian government has provided strong support toward digital transformation. Various policies, infrastructure, programs, and other initiatives have been implemented to ensure the successful inception of digital society in Malaysia. Internationally, the aims of OER have been comprehensively stated in the Paris OER Declaration 2012 (UNESCO 2012) in which relevant strategies and policies on OER were reinforced. So far, many governments and institutions of higher learning (IHLs) have already implemented policies for promoting OER. For instance, many universities (e.g., University of Edinburgh in UK and Queensland University of Technology in Australia) have developed policies that encourage academic staff and students to fully utilize OER for educational purposes (Stevens, Bradbury, and Hutley 2017).

Realizing the importance of OER, the Ministry of Higher Education, Malaysia, has also proposed the adoption and development of open courseware through national MOOCs in local IHLs under the Malaysian Education Blueprint (Higher Education) 2015–2025 (MOE Malaysia 2015). In addition, Malaysian Qualifications Agency (MQA, 2017) has revised the Malaysia Qualifications Framework (MQF) and produced MQF version 2.0. The revised MQF consists of 5 clusters of learning outcomes, with functional work skills as one of the important ones that is focused on digital skills. Gradually, IHLs are expected to revise their program curriculum with embedded digital skills to comply with the requirements of MQF 2.0.

Besides, there is a government-based agency named Malaysia Digital Economy Corporation (MDEC). As a primary institution that pushes for digital economy transformation in Malaysia, MDEC provides training courses, certification programs, mentoring, employment tools, and so forth for upskilling Malaysians to close the digital gaps with industry's needs. For instance, MDEC has partnered with Coursera.org, an

OER platform, to provide certification courses for partnership with selected strategic partners like JobStreet.com, Coursera, HELP University, and other Premier Digital Tech Institutions. Two main platforms created by MDEC are #MyDigitalWorkforce Jobs Platform[13] and Digital Skills Training Directory,[14] to improve Malaysian digital competencies and embrace the IR 4.0.

1.5 CONCLUSION

This chapter has reviewed the recent progress and development in digital competency, open educational resources (OER), and open-source software (OSS) and discussed how OSS and OER can be beneficial toward achieving digital competency. Overall, digital competency is constantly evolving due to its dynamic nature and progress in ICT. Due to the changing ICT industry landscape toward adopting OSS and OER, governments, EIs, and educators need to rethink their strategies in the digital economy. In that regard, the OER and OSS provides a feasible solution toward the sustainable development of digital competency. Future research in this area shall explore the research issues of adopting OER and OSS in digital competency education from multiple focuses such as role, policies, strategies, and technology involved. For instance, how educators adapt to their new role as OER indexers, designers, and creators; adaptive or personalized curriculum design for learners; and technologies involved in aiding the aforementioned processes. All these are areas await further exploration by the research community.

REFERENCES

APEC. 2020. Closing the Digital Skills Gap Report: Trends and Insights Perspectives on. Singapore: APEC Human Resources Development Working Group.

Atenas, J., & Havemann, L. 2013. Quality assurance in the open: An evaluation of OER repositories. *The International Journal for Innovation and Quality in Learning*, *1*(2), 22–34.

Bass, D., & Newcomer, E. 2018. Microsoft Buys GitHub for $7.5 Billion, Going Back to Its Roots. *Bloomberg*. https://www.bloomberg.com/news/articles/2018-06-04/microsoft-agrees-to-buy-coding-site-github-for-7-5-billion.

Bliss, T.J., & Smith, M. 2017. A brief history of open educational resources. *Open: The philosophy and practices that are revolutionizing education and science*:9–27.

Brolpito, A. 2018. Digital Skills and Competence, and Digital and Online Learning. *European Training Foundation*.

Coursera. 2020. Global Skills Index 2020. https://pages.coursera-for-business.org/rs/748-MIV-116/images/gsi2020_final.pdf.

Crompton, H. 2017. *ISTE Standards for Educators: A Guide for Teachers and Other Professionals*. Eugene: International Society for Technology in Education.

Education and Training Foundation, ETF. 2018. Taking Learning to the Next Level: Digital Teaching Professional Framework Full Reference Guide. https://www.etfoundation.co.uk/dtpf.

Gho, C.Y. 2019. The Hays Global Skills Index 2019/20. https://www.hays-index.com/core/uploads/2019/09/HAYS_GLOBAL_SKILLS_INDEX_2019-20_FINAL.pdf.

[13] MDEC #MyDigitalWorkforce Platform: https://mdec.my/digital-economy-initiatives/for-the-people/-mydigitalworkforce-jobs/

[14] MDEC Digital Skills Training Directory: https://mdec.my/digitalskillstrainingdirectory/

Hepburn, G., & Buley, J. 2006. Getting open source software into schools: Strategies and challenges. *Innovate: Journal of Online Education, 3*(1), 6.

Hine, P. 2011. UNESCO ICT competency framework for teachers. Paris: United Nations Educational.

JobStreet.com Malaysia. 2019. Job Outlook 2019. https://www.jobstreet.com.my/en/cms/employer/employers-focused-on-hiring-talent-with-digital-skills/.

Khan, M. A., & UrRehman, F. 2012. Free and open source software: Evolution, benefits and characteristics. *International Journal of Emerging Trends & technology in Computer Science,* 1(3), 1–7.

Lakhan, S.E., & Jhunjhunwala, K. 2008. Open source software in education. *Educause Quarterly,* 31(2), 32. https://er.educause.edu/articles/2008/5/open-source-software-in-education.

Malaysia Qualifications Agency. 2017. *Malaysian Qualifications Framework (MQF)* (2nd ed.) Malaysia: Malaysia Qualifications Agency (MQA).

McAndrew, P., Farrow, R., Law, P., & Elliott-Cirigottis, G. 2012. Learning the lessons of openness. *Journal of Interactive Media in Education.* http://doi.org/10.5334/2012-10.

Ministry of Education and Higher Education Quebec. 2019. Digital Competency Framework. Quebec, Canada.

MIT Media Lab. 2020. Learning about artificial intelligence: A hub of MIT resources for K-12 students. *MIT News.* https://news.mit.edu/2020/learning-about-artificial-intelligence-hub-of-mit-resources-k-12-students-0407.

MOE Malaysia. 2015. Blueprint 2015–2025 (Higher Education). Putrajaya, Malaysia: Ministry of Education.

National Institute of Educational Technologies and Teacher Training. 2017. Digital Competence Framework for Teachers (CDCFT 10/2017). Spain: National Institute of Educational Technologies and Teacher Training.

Navarrete, R., Luján-Mora, S., & Peñafiel, M. 2016. Use of open educational resources in E-learning for higher education. *2016 Third International Conference on eDemocracy & eGovernment (ICEDEG).*

Park, Y. 2019. DQ Global Standards Report 2019 Common Framework for Digital Literacy, Skills and Readiness; Report. *DQ Institute.*

Sawers, P. 2021. Microsoft: Open source is now the accepted model for cross-company collaboration. https://venturebeat.com/2021/01/14/microsoft-open-source-is-now-the-accepted-model-for-cross-company-collaboration/.

Spante, M., Hashemi, S. S., Lundin, M., & Algers, A. 2018. Digital competence and digital literacy in higher education research: Systematic review of concept use. *Cogent Education, 5* (1):1519143. doi:10.1080/2331186x.2018.1519143.

Stevens, J., Bradbury, S., & Hutley, S. 2017. Open education in practice–how policy can lead to positive change. *Journal of the Australian Library and Information Association, 66*(3), 249–258.

Thankachan, B., & Moore, D. R. 2017. Challenges of implementing free and open source software (FOSS): Evidence from the indian educational setting. *The International Review of Research in Open and Distributed Learning, 18*(6). doi:10.19173/irrodl.v18i6.2781.

UNESCO. 2012. Paris OER Declaration, 2012 World Open Educational Resources (OER) Congress. UNESCO, Paris.

Wei, Z. 2011. Research on the application of Open Source Software in digital library. *Procedia Engineering, 15*, 1662–1667.

Zainudin, F. 2016. Developing Digital Competency Standards (DCS) as a Tool to Measure Students' Digital Skills. *36th APEC e-Learning Training Program*, Seoul, Korea, 10–19 July 2016.

2 Virtual Reality for Enhancing Student Teachers' Science Competencies

Hamid Ikram, Muhammad Ayub Buzdar, and Sumera Hanif
Government College University Faisalabad

CONTENTS

DOI: 10.1201/9781003143796-2

2.1 INTRODUCTION

Virtual Reality (VR) as a transformative technological tool offers the experience of alternative reality by stimulating users' senses, including sight, hearing, and touch (Pope, 2018). According to Singh et al. (2020), "virtual reality is developing simulated expertise which is somewhat similar to the real-time situation" (p. 661). The foundation of VR is located in the panoramic paintings of the nineteenth century; however, the name "Virtual Reality" was first used in 1987 by Jaron Lanier—founder of the Visual Programming Lab (Virtual Reality Society, 2017).

The VR systems are categorized as nonimmersive, semi-immersive, and immersive based on their ability to transform the perception of the real world with the virtual one (Di Natale et al., 2020). Immersion is the involvement of a person in the virtual environment. Both high-end and low-budget VR head-mounted displays (HMDs) attract users by their immersion and visualization properties (Radianti et al., 2020). Many popular tech companies including Oculus, Google, Sony, Samsung, HTC, and Facebook have launched distinctive virtual reality platforms that are versatile in their immersion (Di Natale et al., 2020), cost, and portability (Hurrell & Baker, 2020). Oculus, which was later purchased by Facebook, launched the first-ever high-end VR headset named "Rift." Later, Facebook introduced other self-contained and inexpensive HMDs called Oculus Go and Quest. Moreover, the HTC Vive is also a high-end, fully equipped, and sophisticated device that provides a better VR experience to its consumers (Radianti et al., 2020). HTC has recently released a self-contained VR headset known as Vice Cosmos (Hurrell & Baker, 2020). The Google Cardboard is a popular low-cost portable headset that works with the smartphone and enables everyone to experience an immersive virtual environment (Lee et al., 2017; Radianti et al., 2020).

The teacher education programs of most public sector universities in Pakistan are less likely to produce progressive educators. The teacher training in most developing countries like Pakistan focuses on the theoretical approach, and rarely follows practical applications of modern teaching tools (Ashraf et al., 2005). Generally, the teacher training programs in Pakistan are less practical and rely on traditional methods. Therefore, the trainees are not exposed to newer ideas and innovative pedagogies during their teacher preparation programs (Davies & Iqbal, 1997). The major barrier for the evolution of student teachers to become progressive teachers in the Pakistani context is reluctance in adopting newer pedagogical techniques. Due to the dominance of the traditional teacher training practices, the transition from prospective teacher to classroom instructor stays conventional. As a result, these dull practices minimize the outcome of the teacher training and halt its impact to improve the quality of the education process (Westbrook et al., 2009).

The present study aimed to bring innovation in the conventional practices of teacher education by introducing science learning with the help of VR. There is sufficient empirical evidence that promotes the implementation of VR for teachers' long-term professional development (Stavroulia et al., 2019). The incorporation of smart technologies such as VR HMDs could enhance the quality of teacher's work and has the potential to become a useful tool in teacher training. The teachers and student teachers must adopt emerging technologies as these tools are more user-friendly, accessible, and affordable (Fransson et al., 2020). The smartphone-supported Google Cardboard VR experience is possible to incorporate into a low-resource educational setting (Šašinka et al., 2019).

2.2 LITERATURE REVIEW

2.2.1 VR in the Education Domain

The conventional educational tools have evolved from pens, pencils, and books to more interactive learning technologies that aid in the effective delivery of content to learners (Alfalah, 2018). VR is one such interactive tool (Shim et al., 2003) that effectively supplements both the teaching and learning process and revolutionizes the whole education system. The innovative technology, particularly VR, has the potential to reform education by ensuring access and engaging learners (Gadelha, 2018). At present, VR has developed rapidly in many fields of education, ranging from space exploration (Christensen & Knezek, 2020) to medical education (Singh et al., 2020), from business education (Lee et al., 2017) to engineering education (Alhalabi, 2016).

A VR-enriched learning environment stimulates learners to interact and construct new knowledge and skills by verifying, enhancing, or invalidating their previous knowledge (Lamb & Etopio, 2020). Keller et al. (2018) claimed, "VR is compatible with the learning theory of constructivism through the possibility of immersion, interaction, 3D representation and the possibility of adopting several perspectives" (p. 53).

The VR platforms have transformed the entire education process (Cooper et al., 2019) as it has been proved to be an efficient way to deliver the learning material through experiential and learner-centered approaches (Hurrell & Baker, 2020). The other key benefit of VR immersion in education includes the first-hand experience of unobservable phenomena of the real world, which is not possible otherwise, by triggering learners' motivation and engagement through the situated-learning approach (Di Natale et al., 2020). Mantovani (2001) suggested, "virtual environments can provide a rich, interactive, engaging educational context, supporting experiential learning" (p. 208). The learners not only enjoy VR-supported instruction but also feel motivated toward the subject matter (Lee et al., 2017). The VR learning media positively affects students' beliefs, and it has enormous potential for students' success, learning outcome, attention, and collaboration. This emerging technology is getting wide attention to reinvent and reconstruct the instructional process at all levels (Zhang et al., 2020).

2.2.2 VR for Science Instruction

Innovative teaching technologies such as VR have contributed to the field of science education (Artun et al., 2020). VR as an instructional tool is effective to teach all subjects of science including biology, physics, chemistry, and earth science. The students can get hands-on experience of abstract scientific concepts in a virtual environment that inspires deep learning (Shim et al., 2003). With the help of VR platforms, the unobservable ideas of science can be visualized in a 3D style and successfully presented to learners (Alfalah, 2018; Artun et al., 2020). As a result, the student's achievement scores in science significantly improve when they are taught with VR as compared to that with 2D learning media (Shim et al., 2003). There are other benefits of teaching science with the help of VR. First, VR can stimulate multi-sensory organs of learners (Pope, 2018) as well as enhance students' motivation and interest in the subject matter (Lee et al., 2017; Shim et al., 2003). Second, the VR-enabled learning activities provide a real-time interface between the user and the technology (Mantovani, 2001; Shim et al., 2003). Third, VR provides a safe, effective, and controllable environment to conduct science experiments, which are expensive or too dangerous to perform in traditional settings. Finally, students have more freedom and take responsibility for their learning in virtual settings (Shim et al., 2003).

2.2.3 VR in Higher Education Settings

Recently, the use of VR is growing in the tertiary education sector (Hagge, 2020) due to its affordability and capability to offer a 3D experience to learners (Bower et al., 2020). From past research findings, the significance of VR is evident in various university fields including business education (Lee et al., 2017), engineering education (Alhalabi, 2016), and electrical/electronics technology (Ogbuanya & Onele, 2018). Since the launch of VR in education, the institutions of higher learning have been revamping to digitize their resources for students' optimal experience. Subsequently, the universities are handing VR learning media to their students (Alfalah, 2018). One such example is seen at the University of Calgary where a VR studio, located in Tylor Family Digital Library, is providing experiential learning to the university students. This advanced VR studio facilitates undergrad students belonging to diverse majors including dance, literature, art history, and kinesiology—through virtual field trips, apps development, and language learning (Hurrell & Baker, 2020). The university students do not only achieve better grades with the help of VR but also show higher engagement, attentiveness, and interest in their learning activities. The interactive nature of VR enhances university students' engagement; likewise, a meaningful learning experience increases their interest (Ogbuanya & Onele, 2018).

Many developing countries lack VR facilities for their university students (Ogbuanya & Onele, 2018) due to the cost associated with its adoption in higher education settings (Alfalah, 2018). The findings of a meta-analysis showed that 76% of studies, conducted in higher education settings, used expensive HMDs such as Oculus Rift and HTC Vive as compared to inexpensive mobile-based VR such as Google Cardboard (Radianti et al., 2020). However, Papachristos et al. (2017) did not find any significant difference in usability, satisfaction, and academic benefits between expensive headsets and inexpensive smartphone-enabled VR headsets. The researchers concluded that mobile-based VR is equally capable to provide pedagogical support and interaction in virtual learning environments. Other factors that hinder university faculty to bring VR into their classrooms include lack of facilities, insufficient technical support, and cultural barriers (Alhudaithy, 2019).

2.2.4 VR and Teachers' Training

Since VR has entered the arena of teaching as an instructional tool, the real beneficiaries are students (Alfalah, 2018). Gadelha (2018) said, "VR has the potential to dramatically shift how teachers teach and, more importantly, how learners learn" (p. 41). This immersive condition is fascinating for both teachers and students because of such an enjoyable and authentic learning experience (Alfalah, 2018) in which students and teachers can interact, engage, and collaborate (Hurrell & Baker, 2020). So, to give such learning experiences to students and teaching expertise to teachers, the teachers' professional development in using VR is essential (Alfalah, 2018). A rigorous teachers' training program that conceives VR may result in better teaching practices and, hence, student learning (Bower et al., 2020).

The teacher's ability to integrate VR into instructional practices could transform the traditional classroom into a technology-oriented learning space (Alfalah, 2018). Many teachers view VR as an effective tool for teaching, and the courses such as science, history, and social study are the top subjects to teach with VR (Zaino, 2016). The VR-enabled environment has the potential to train preservice science teachers and help them to acquire scientific skills (Artun et al., 2020). However, technology comes with a cost, but there is an affordable way for teachers to start with VR by using inexpensive headsets that work with the smartphones, such as Google Cardboard (Zaino, 2016).

2.3 THE CURRENT STUDY

This study aimed to introduce reforms in preservice teacher education with the help of transformative technology. The study strived to bring innovation in conventional practices and introduced science learning with VR. The research objectives included teaching and training the prospective science teachers with affordable VR headsets such as Google Cardboard and YouTube 360° learning videos.

2.4 MATERIAL AND METHODS

2.4.1 Research Design

The nature of the study was experimental, and it followed a pretest and posttest control group design. The experimental group received the intervention, while the control group remained unchanged. The control and experimental groups were pre- and posttested before and after the intervention.

2.4.2 Participants

The participants of this study were 27 student teachers enrolled in a B.Ed. (Hons.) science class in a public sector university of the Punjab province of Pakistan. The participants were randomly divided into two groups through a simple random sampling technique. 14 students (six male and eight female) were placed in the experimental group (VR group) and 13 students (five male and eight female) were placed in the control group (non-VR group). The participants were aged between 20 and 24 years ($M_E = 20.86$, $M_C = 21.38$). The average GPA of the participants during the previous semester was ($M_E = 3.59$, $M_C = 3.23$). Twelve students resided in rural areas (six in each treatment and control group), and 15 students belonged to urban areas (eight in the intervention group and seven in the traditional group). In the VR group, two students hailed from the academic background of arts, while 12 students belonged to science. In the non-VR group, four students were from the arts background and nine students belonged to the science discipline (Table 2.1).

2.4.3 Homogeneity of Groups

The baseline test (pretest) showed that the participants in both VR and non-VR groups were homogeneous concerning their previous knowledge of the content being taught. The preintervention test scores of both study groups were analyzed to find the significant mean difference. However, the independent-sample *t*-test showed insignificant results as $p > 0.05$. Table 2.2 indicates pretest scores of non-VR group ($M = 5.53$, $SD = 2.40$) and VR group ($M = 4.57$, $SD = 3.15$) in the solar unit; pretest scores of non-VR group ($M = 2.46$, $SD = 1.20$) and VR group ($M = 2.78$, $SD = 1.05$) in the energy unit; and total combined baseline scores of non-VR group ($M = 8.00$, $SD = 3.14$) and VR group ($M = 7.36$, $SD = 3.65$), tested at the alpha $= 0.05$ significance level.

TABLE 2.1
Participants' Demographics

Variable		Experimental Group	Control Group
Age (*M*)		20.86	21.38
GPA (*M*)		3.59	3.23
Gender	Male	6 (43%)	5 (39%)
	Female	8 (57%)	8 (61%)
Residence	Rural	6 (43%)	6 (46%)
	Urban	8 (57%)	7 (54%)
Academic background	Arts	2 (14%)	4 (31%)
	Science	12 (86%)	(69%)

TABLE 2.2
Mean, Standard Deviation, and Growth in Students' Learning in Experimental and Control Groups

Group	Content	Max Score in Pretest	Students Scored in Pretest: *M* and (*SD*)	Max Score in Posttest	Students Scored in Posttest: *M* and (*SD*)	Growth (%)
Experimental	Solar	14	4.57 (3.15)	13	10.79 (3.01)	340
	Energy	8	2.78 (1.05)	9	5.71 (2.02)	132
	Combine	22	7.36 (3.65)	22	16.50 (4.74)	198
Control	Solar	14	5.53 (2.40)	13	7.31 (2.02)	79
	Energy	8	2.46 (1.20)	9	3.85 (1.07)	76
	Combine	22	8.00 (3.14)	22	11.15 (1.34)	58

2.4.4 Data Collection Instrument

The researchers developed the preintervention and postintervention tests to collect data for this inquiry. Both tests included the items from the units of energy and solar system—these two units were only taught to both groups during this study. The researchers developed as well as adopted the test items from the science textbook. The pretest had of total 22 items out of which 14 items focused on the solar system and the remaining 8 covered the unit of energy. The posttest test also had 22 items, but 13 items covered the unit of the solar system and 9 items tested the concepts of energy. Furthermore, both tests included four parts. The first part of both tests included ten multiple-choice items; the second part of both tests included four true/false statements, the third part included four questions related to filling in the blanks, and the fourth part of both tests was a four-item matching column. The scoring procedure of the test items was 0 and 1, where 0 was a wrong answer and 1 was a correct answer. Thus, the participants' total achievement was measured out of total of 22 scores in both pre- and posttest.

2.4.5 Validity and Reliability

The experts' opinion was taken for content validity of both tests. The two subject experts were requested to review the test items regarding their representation, difficulty, structure, and appropriateness of the covered content. The experts suggested removing three items because of inconsistency and adding two items to enhance the scope of the tool. Also, five items were rephrased and restructured for the understanding of the test takers. Thus, the final form of both pretest and posttest included 22 items.

Also, the Cronbach alpha reliability values of pretest and posttest were .70 and .80, respectively.

2.5 PROCEDURE

This study was completed in two phases.

2.5.1 First Phase: Lesson Planning and Material Development

During this phase, the material development such as lesson planning, content selection, tech acquisition, and installation was completed. The researchers were the course instructors for this science class that was offered to the preservice teachers in the spring semester. Two units, i.e., solar system and energy, were selected from the syllabus of the science class to teach with and without VR. Subsequently, the researchers developed VR-assisted lessons to be delivered to the treatment group. The major equipment for this study included the Google Cardboard VR headsets, smartphones, audio handsfree, Wi-Fi, and 3D VR 360° videos related to the content—all videos were selected from the YouTube platform. Both groups were pretested on the above-stated science concepts during this phase as well.

2.5.2 Second Phase: Experimental Stage—Treatment for One Group and No Treatment for Second Group

This study phase was completed as the semester concluded. The instructors who taught the entire science course to both study groups taught the same content with and without VR during alternative hours. During the formal delivery of the lessons in the class, the treatment group was provided with the opportunities to watch 3D 360° VR videos of varying lengths, with the help of Head-Mounted Displays (Google Cardboards). In contrast, the control group received no treatment and learned the science concepts through their conventional methods. The traditional method included lectures aided with textbooks, static 2D pictures, and whiteboards. After the semester-long treatment, both groups were posttested.

2.6 RESULTS

2.6.1 Descriptive Analysis

The descriptive statistics, including mean and standard deviations, were calculated for the participants' achievement scores in both pre- and posttest. The groups (intervention and nonintervention) and content (solar, energy, and combined) distribution of the students' achieved scores with percentage growth is provided in Table 2.2.

2.6.2 Comparison between Participants' Science Scores

The comparison of the preintervention and postintervention science scores of the participants (VR and non-VR groups) was conducted with the help of a paired-sample *t*-test. The analysis in the table below shows a significant mean difference between all the domains of pre- and posttest scores of participants in the VR group. The prospective teachers in the non-VR group, however, showed a significant mean difference in the pre- and posttest scores in the energy unit and combined contents. The mean differences showed variances in the growth of students' scores in both groups as well as in different contents (Table 2.3).

TABLE 2.3
Comparison of Participants' Preintervention and Postintervention Scores in VR and Non-VR Groups (Paired Sample *t*-Test)

Group	Content	*MD*	*t*	*df*	*p*
Experimental	Solar	6.21	4.70	13	0.00
	Energy	2.93	5.65	13	0.00
	Combine	9.14	5.86	13	0.00
Control	Solar	1.77	1.91	12	0.08
	Energy	1.38	3.96	12	0.00
	Combine	3.15	3.33	12	0.01

$p < 0.05$

TABLE 2.4
Comparison of Participants' Science Growth in VR and Non-VR Groups (Independent Sample *t*-Test)

Content	Experimental Growth (*M*)	Control Growth (*M*)	*MD*	*t*	*df*	*p*
Solar	3.40	0.79	2.61	2.24	25	0.03
Energy	1.32	0.76	0.56	1.36	25	0.19
Combine	1.98	0.58	1.40	2.22	25	0.04

$p < 0.05$

2.6.3 Comparison between the Growth in Participants' Science Scores

The comparison in science growth index between the participants of the VR and non-VR groups was conducted via an independent-sample *t*-test. The results shown in Table 2.4 reveal that the increase in the participants' scores in the solar unit and combined content is significantly higher in the VR group than that in the non-VR group. However, the participants did not show a significant mean difference in growth in the energy unit.

2.7 DISCUSSION AND CONCLUSION

The learning experiences presented with the help of transformative technologies such as VR are interactive, collaborative, and keep the students engaged. In contrast, the students remain passive in the conventional classroom settings, and their interest in lessons is persistently low due to the non-tech study environment. VR is a kind of an innovative technological tool that can enhance learning by stimulating, engaging, and motivating the students of all fields (Ogbuanya & Onele, 2018).

The present inquiry revealed that a VR-enriched experience positively enhanced student teachers' science learning. The participants, enrolled in an undergraduate teacher education program, scored significantly higher in a VR-based learning environment as compared to their peers in traditional settings. The results were consistent with the findings of Artun et al. (2020) who claimed that VR technology is beneficial for training prospective science teachers to develop scientific skills. Their inquiry revealed significant improvement in the science scores after VR-intervention among prospective science teachers in Turkey.

The present study uncovered that the semester-long VR-enabled learning experiences exponentially enhanced the overall science achievement of the treatment group. These findings resonate with a quasi-experimental study conducted by Liu et al. (2020) who discovered that the experimental group, who received VR-based science lessons with the help of head-mounted displays, significantly scored higher as compared to the nonintervention group who learned the same content with routine instructional practices. Moreover, Huang et al. (2019) found that VR is more immersive and engaging for science-knowledge retention as compared to other technology.

Another key finding of this inquiry asserted that the participants who received VR intervention particularly, scored higher in the unit of the solar system. These findings correlate with those of Barab et al. (2000) who found that VR is significantly helpful for undergraduate university students to understand various astronomical processes including the solar system. Christensen and Knezek (2020) affirmed that the transformative technology of VR not only scaffolds the learning of the solar system but also engages students in the overall learning process. A meta-analysis conducted by Radianti et al. (2020) revealed that astronomy was one of the popular subjects taught with VR in the higher education domain.

It was observed in this study that VR had a significantly stronger effect on learning the content of the solar system as compared to learning the energy concepts. An underlying reason for such an issue may be the users' acquaintance with VR technology at the beginning of the experiment. As the unit of energy was taught at the beginning of the intervention, the participants might spend more time exploring the VR technology itself and invest less time understanding the content of energy. Previous research strengthens the need for pretraining for users to get familiar with the novel technology like VR, before its actual use to teach the content (Meyer et al., 2019; Moro et al., 2017; Stepan et al., 2017).

This study was delimited to the 27 Pakistani prospective teachers enrolled in a science class of an undergraduate teacher education program. The content taught with the help of VR only included the solar system and energy units. This study is a torchbearer for future studies that intend to see the effect of VR in different settings.

2.8 IMPLICATIONS

This study supports the academic use of affordable mobile-based VR technology, like Google Cardboard, in developing countries like Pakistan. The study has implications for the development of preservice teachers' instructional skills to bring VR into their future classrooms. According to Lamb and Etopio (2020), preservice teachers considered VR as assistance in developing their teaching skills after their very own experience of dealing with VR. The preservice teachers could optimize their future classroom practices by implementing immersive VR (Bower et al., 2020)

In the beginning, technical difficulties and the high cost of VR tools slowed the incorporation of this transformative technology in the education field. But, currently, the advantages of this hands-on immersive experience are being utilized at all levels of education (Nemec et al., 2017). However, there is a huge gap between the technical specification and users' experience of expensive HMDs such as Oculus Rift and inexpensive smartphone-supported VR tools like Google Cardboard. However, there is sufficient empirical evidence to show that the smartphone-based VR platforms could equally contribute to the pedagogical application of VR technology and also provide an acceptable quality of immersion and user experience (Nemec et al., 2017; Papachristos et al., 2017).

The incorporation of VR in education is feasible as well as affordable, even for developing countries like Pakistan. The investigators of this study used Google headsets, smartphones, and VR 360° 3D videos from YouTube. The use of a basic Google headset made with cardboard and a smartphone is viable to provide a VR experience to all learners. The YouTube platform provides many 3D 360° VR videos. Also, several apps on the Google play store provide downloadable content to users. Besides YouTube VR videos, some popular VR-enabled apps to teach about the solar system are Titans of Space, Access Mars, Space VR, and Edge of Home (Christensen & Knezek, 2020).

VR is a helpful instructional tool to deliver diverse science content as the outcomes of this study showed its immense capability to effectively supplement science lessons. Hence, the comparison of the present findings with existing literature is limited because of the scarcity of empirical data on the effects of VR-supported instruction in science classrooms, especially for preservice science teachers. There is still a great need to conduct further research on all potential effects of VR in the education sector. The findings of this inquiry are helpful for the future endeavors of integrating VR technologies in the teaching and learning of science, especially for beginner teachers.

Another avenue that needs exploring is the effect of a VR-enabled learning environment in the time of pandemics like coronavirus disease. According to Carruth (2017), the VR learning environment is safe, controlled, and risk-free. As the VR platforms could provide remote learning (Shim et al., 2003) and reduce face-to-face interaction among individuals (Singh et al., 2020), VR platform can effectively deliver online and hybrid education.

APPENDIX A: CONTENT COVERED FOR ENERGY UNIT

Unit # 1 Energy	Projected Learning Outcomes	Resources for VR Group (Experimental Settings)	Resources for non-VR Group (Traditional Settings)
Types of energy Kinetic energy Potential energy Light energy Sound energy Gravitational energy Heat energy Heat transfer Energy transfer and transformation Law of conservation of energy	Explore different forms of energy, with examples Compare and contrast all types of energy Find out how energy is produced Examine how energy transfer from one medium to another Explore conduction, convection, and radiation with examples Investigate the changes that transform energy from one form to another Understand how energy is converted from one form to another Apply the law into practice	Google Cardboard Headsets Smartphones Internet Audio Handsfree YouTube VR 3D 360° Videos related to Energy	Textbook Whiteboard 2D Pictures

APPENDIX B: CONTENT COVERED FOR SOLAR SYSTEM UNIT

Unit # 2 Solar System	Projected Learning Outcomes	Resources for VR Group (Experimental Settings)	Resources for non-VR Group (Traditional Settings)
Solar system Planets Mercury Venus Earth Mars Jupiter Saturn Uranus Neptune	Explore the size of the Sun and distance of planets from the Sun Find and compare the size of the planets Examine the different characteristics of planets including their surface temperature, orbits, moon, movement, etc. Examine the origin of the earth and its evolution Explore the possibility of life on Mars Be able to recognize each planet by its unique and identifiable features	Google Cardboard Headsets Smartphones Internet Audio Handsfree YouTube VR 3D 360° Videos related to Solar system	Textbook Whiteboard 2D Pictures Globe

REFERENCES

Alfalah, S. F. M. (2018). Perceptions toward adopting virtual reality as a teaching aid in information technology. *Education and Information Technologies*, *23*(6), 2633–2653. https://doi.org/10.1007/s10639-018-9734-2.

Alhalabi, W. (2016). Virtual reality systems enhance students' achievements in engineering education. *Behaviour & Information Technology*, *35*(11), 919–925. https://doi.org/10.1080/0144929X.2016.1212931.

Alhudaithy, H. (2019). Virtual reality from the perspective of Saudi faculty. In F. Meunier, J. Van de Vyver, L. Bradley, & S. Thouësny (Eds.). *CALL and complexity – short papers from EUROCALL 2019* (1st ed., pp. 1–6). Research-publishing.net. https://doi.org/10.14705/rpnet.2019.38.977.

Artun, H., Durukan, A., & Temur, A. (2020). Effects of virtual reality enriched science laboratory activities on pre-service science teachers' science process skills. *Education and Information Technologies*, *25*(6), 5477–5498. https://doi.org/10.1007/s10639-020-10220-5.

Ashraf, D., Khaki, J.-A., Shamatov, D. A., Tajik, M. A., & Vazir, N. (2005). Reconceptualization of teacher education: Experiences from the context of a multicultural developing country. *Journal of Transformative Education*, 3(3), 271–288. https://doi.org/10.1177/1541344605275230.

Barab, S. A., Hay, K. E., Squire, K., Barnett, M., Schmidt, R., Karrigan, K., Yamagata-Lynch, L., & Johnson, C. (2000). Virtual solar system project: Learning through a technology-rich, inquiry-based, participatory learning environment. *Journal of Science Education and Technology*, *9*(1), 7–25. https://doi.org/10.1023/A:1009416822783.

Bower, M., DeWitt, D., & Lai, J. W. M. (2020). Reasons associated with preservice teachers' intention to use immersive virtual reality in education. *British Journal of Educational Technology*, *51*(6), 2215–2233. https://doi.org/10.1111/bjet.13009.

Carruth, D. W. (2017). Virtual reality for education and workforce training. *2017 15th International Conference on Emerging ELearning Technologies and Applications (ICETA).* https://doi.org/10.1109/ICETA.2017.8102472.

Christensen, R., & Knezek, G. (2020). Connecting students to Mars and stars using innovative technologies. *Journal of Educational Multimedia and Hypermedia, 29*(3), 181–192.

Cooper, G., Park, H., Nasr, Z., Thong, L. P., & Johnson, R. (2019). Using virtual reality in the classroom: Preservice teachers' perceptions of its use as a teaching and learning tool. *Educational Media International, 56*(1), 1–13. https://doi.org/10.1080/09523987.2019.1583461.

Davies, L., & Iqbal, Z. (1997). Tensions in teacher training for school effectiveness: The case of Pakistan*. *School Effectiveness and School Improvement*, 8(2), 254–266. https://doi.org/10.1080/0924345970080205.

Di Natale, A. F., Repetto, C., Riva, G., & Villani, D. (2020). Immersive virtual reality in K-12 and higher education: A 10-year systematic review of empirical research. *British Journal of Educational Technology, 51*(6), 2006–2033. https://doi.org/10.1111/bjet.13030.

Fransson, G., Holmberg, J., & Westelius, C. (2020). The challenges of using head mounted virtual reality in K-12 schools from a teacher perspective. *Education and Information Technologies, 25*(4), 3383–3404. https://doi.org/10.1007/s10639-020-10119-1.

Gadelha, R. (2018). Revolutionizing education: The promise of virtual reality. *Childhood Education, 94*(1), 40–43. https://doi.org/10.1080/00094056.2018.1420362.

Hagge, P. (2020). Student perceptions of semester-long in-class virtual reality: Effectively using "Google Earth VR" in a higher education classroom. *Journal of Geography in Higher Education*, 1–19. https://doi.org/10.1080/03098265.2020.1827376.

Huang, K.-T., Ball, C., Francis, J., Ratan, R., Boumis, J., & Fordham, J. (2019). Augmented versus virtual reality in education: An exploratory study examining science knowledge retention when using augmented reality/virtual reality mobile applications. *Cyberpsychology, Behavior, and Social Networking, 22*(2), 105–110. https://doi.org/10.1089/cyber.2018.0150.

Hurrell, C., & Baker, J. (2020). Immersive learning: Applications of virtual reality for undergraduate education. *College & Undergraduate Libraries.* https://doi.org/10.1080/10691316.2020.1796879.

Keller, T., Glauser, P., Ebert, N., & Brucker-Kley, E. (2018). Virtual Reality at secondary school – First results. *15th International Conference on Cognition and Exploratory Learning in Digital Age (CELDA 2018)*, 53–60.

Lamb, R., & Etopio, E. A. (2020). Virtual Reality: A tool for preservice science teachers to put theory into practice. *Journal of Science Education and Technology, 29*(4), 573–585. https://doi.org/10.1007/s10956-020-09837-5

Lee, S. H. (Mark), Sergueeva, K., Catangui, M., & Kandaurova, M. (2017). Assessing Google Cardboard virtual reality as a content delivery system in business classrooms. *Journal of Education for Business, 92*(4), 153–160. https://doi.org/10.1080/08832323.2017.1308308.

Liu, R., Wang, L., Lei, J., Wang, Q., & Ren, Y. (2020). Effects of an immersive virtual reality-based classroom on students' learning performance in science lessons. *British Journal of Educational Technology, 51*(6), 2034–2049. https://doi.org/10.1111/bjet.13028

Mantovani, F. (2001). VR learning: Potential and challenges for the use of 3D environments in education and training. In: Riva, G. & Galimberti, C. (Eds.). *Towards Cyber Psychology: Mind, Cognitions and Society in the Internet Age* (pp. 207–226). Amsterdam, The Netherlands: IOS Press.

Meyer, O. A., Omdahl, M. K., & Makransky, G. (2019). Investigating the effect of pre-training when learning through immersive virtual reality and video: A media and methods experiment. *Computers & Education, 140*, 103603. https://doi.org/10.1016/j.compedu.2019.103603.

Moro, C., Štromberga, Z., Raikos, A., & Stirling, A. (2017). The effectiveness of virtual and augmented reality in health sciences and medical anatomy: VR and AR in health sciences and medical anatomy. *Anatomical Sciences Education*, *10*(6), 549–559. https://doi.org/10.1002/ase.1696.

Nemec, M., Fasuga, R., Trubac, J., & Kratochvil, J. (2017). Using virtual reality in education. *2017 15th International Conference on Emerging ELearning Technologies and Applications (ICETA)*. https://doi.org/10.1109/ICETA.2017.8102514.

Ogbuanya, T. C., & Onele, N. O. (2018). Investigating the effectiveness of desktop virtual reality for teaching and learning of electrical/electronics technology in universities. *Computers in the Schools*, *35*(3), 226–248. https://doi.org/10.1080/07380569.2018.1492283.

Papachristos, N. M., Vrellis, I., & Mikropoulos, T. A. (2017). A comparison between Oculus Rift and a low-cost smartphone VR headset: Immersive user experience and learning. *2017 IEEE 17th International Conference on Advanced Learning Technologies (ICALT)*, 477–481. https://doi.org/10.1109/ICALT.2017.145.

Pope, H. (2018). Virtual and augmented reality in libraries. *Library Technology Reports*, *54*(-6), 5–23. https://doi.org/10.5860/ltr.54n6.

Radianti, J., Majchrzak, T. A., Fromm, J., & Wohlgenannt, I. (2020). A systematic review of immersive virtual reality applications for higher education: Design elements, lessons learned, and research agenda. *Computers & Education*, *147*, 103778. https://doi.org/10.1016/j.compedu.2019.103778.

Šašinka, Č., Stachoň, Z., Sedlák, M., Chmelík, J., Herman, L., Kubíček, P., Šašinková, A., Doležal, M., Tejkl, H., Urbánek, T., Svatoňová, H., Ugwitz, P., & Juřík, V. (2019). Collaborative immersive virtual environments for education in Geography. *International Journal of Geo-Information*, *8*(3), 1–15. https://doi.org/10.3390/ijgi8010003.

Shim, K.-C., Park, J.-S., Kim, H.-S., Kim, J.-H., Park, Y.-C., & Ryu, H.-I. (2003). Application of virtual reality technology in biology education. *Journal of Biological Education*, *37*(2), 71–74. https://doi.org/10.1080/00219266.2003.9655854.

Singh, R. P., Javaid, M., Kataria, R., Tyagi, M., Haleem, A., & Suman, R. (2020). Significant applications of virtual reality for COVID-19 pandemic. *Diabetes & Metabolic Syndrome: Clinical Research & Reviews*, *14*(4), 661–664. https://doi.org/10.1016/j.dsx.2020.05.011.

Stavroulia, K. E., Christofi, M., Baka, E., Michael-Grigoriou, D., Magnenat-Thalmann, N., & Lanitis, A. (2019). Assessing the emotional impact of virtual reality-based teacher training. *The International Journal of Information and Learning Technology*, *36*(3), 192–217. https://doi.org/10.1108/IJILT-11-2018-0127.

Stepan, K., Zeiger, J., Hanchuk, S., Del Signore, A., Shrivastava, R., Govindaraj, S., & Iloreta, A. (2017). Immersive virtual reality as a teaching tool for neuroanatomy: Immersive VR as a neuroanatomy teaching tool. *International Forum of Allergy & Rhinology*, *7*(10), 1006–1013. https://doi.org/10.1002/alr.21986.

Virtual Reality Society. (2017). *History of virtual reality*. Virtual Reality Society. https://www.vrs.org.uk/virtual-reality/history.html.

Westbrook, J., Shah, N., Durrani, N., Tikly, C., Khan, W., & Dunne, M. (2009). Becoming a teacher: Transitions from training to the classroom in the NWFP, Pakistan. *International Journal of Educational Development*, *29*(4), 437–444. https://doi.org/10.1016/j.ijedudev.2008.12.001.

Zaino, J. (2016, June 27). *Teachers ready for virtual reality in education*. Teachers Ready for Virtual Reality in Education. https://insights.samsung.com/2016/06/27/teachers-ready-for-virtual-reality-in-education/.

Zhang, H., Yu, L., Ji, M., Cui, Y., Liu, D., Li, Y., Liu, H., & Wang, Y. (2020). Investigating high school students' perceptions and presences under VR learning environment. *Interactive Learning Environments*, *28*(5), 635–655. https://doi.org/10.1080/10494820.2019.1709211.

3 Learners' Perspective of Emergency Remote Teaching in a Public Higher Education Institution

Dorothy DeWitt, Donnie Adams, and Bambang Sumintono
University of Malaya

CONTENTS

3.1 INTRODUCTION

The coronavirus disease (COVID-19) pandemic required a sudden change in the way higher education institutions operate and in the delivery of instructions coupled with a drastic change in course curricula. The switch to emergency remote teaching was made mandatory rather than optional. Courses initially designed for face-to-face interaction are now shifted online due to an emergency closure of higher education institutions worldwide in an attempt to contain the spread of the COVID-19 pandemic. According to the United Nations Education, Scientific and Cultural Organization (2020), nearly 1.95 billion learners in almost 195 countries are away from universities and colleges.

A forced shift to emergency remote teaching gave lecturers little time to prepare for the transition. They were required to move their courses to an emergency remote

DOI: 10.1201/9781003143796-3

teaching mode, in some cases, within the span of a week. In addition, adjustments to learners' learning outcomes were unavoidable. Conversely, heading into this "new normal" also requires learners to adapt to this change as well. Learners were instructed to attend online class via videoconferencing tools, mostly those preferred by their lecturers, in order to maintain formal learning (Li & Lalani, 2020). Asad et al. (2020) observed that Information and Communications Technology (ICT) tools facilitated educators in their teaching process and made learners' online learning more effective. However, little is known about learner's satisfaction and needs.

Thus, the purpose of this research was to investigate learners' satisfaction, needs, and challenges with emergency remote teaching such as connectivity, devices, and learning strategies in a prominent public higher education institution during the emergency. This study contributes to the existing literature on emergency remote teaching by providing empirical evidence leading to new knowledge in the Malaysian context. The following section provides a literature review on technology in higher education, emergency remote teaching, and transferability and viability of technology use. Next, the methodology and results are presented, followed by discussion of the results. Finally, the chapter will provide insights into the implications of emergency remote teaching, the limitation of this study, and suggestions for future studies.

3.2 LITERATURE REVIEW

3.2.1 Technology in Higher Education

In Malaysia, the Ministry of Higher Education (MOHE) views Globalized Online Learning as a thrust for enhancing higher education where technology-enabled innovations such as video-conferencing and live streaming is employed (MOHE, 2015). Toward the goal of having online learning in higher education for lifelong learning, it was suggested that HEIs implement 70% of the programs in the blended learning mode (MOHE, 2015). All higher education institutions (HEIs) in Malaysia have Learning Management Systems (LMS) and, with the promotion of blended learning (Adams & DeWitt, 2021), the option to go online during the pandemic was considered to be less disruptive and lecturers and leaners in the HEIs could have fewer constraints (Adams et al., 2020a).

Collaborative learning seems to encourage richer learning experiences. As leaners work toward a shared purpose, they collaborate, argue, and defend their point of view as they explore their understandings (DeWitt & Koh, 2020). Making mistakes, debating, and analyzing the information acquired to make conclusions and generate new knowledge in a safe environment enables higher-level thinking processes and can be an impetus for innovation and creativity (DeWitt & Koh, 2020). Many technology tools can enhance the collaborative processes for knowledge generation and are commonly referred to as Web 2.0 tools.

There have been some studies on collaborative learning technology tools in higher education. Wikis (Biasutti, 2017; DeWitt, Alias, Siraj, & Hutagalung, 2014), interactive virtual walls (*Padlet*) (DeWitt & Koh, 2020; Dewitt, Alias, & Siraj, 2015), and discussion forums (Burhan-Horasanlı & Ortaçtepe, 2016); DeWitt, Alias, & Md. Nor, 2016) have the potential to enhance knowledge sharing and knowledge generation and engage learners at the same time due to its interactive nature. On the other hand, there are also resources and e-books delivered on LMS for learners (Noor, Attaran, & Alias, 2015), but if they are not used appropriately, it will be only for the transmission of knowledge.

Before the pandemic, lecturers in HEIs had been using the LMS to ensure there was blended learning (Adams et al., 2020b, 2018). A survey of technology tools used by lecturers in a HEI where blended learning was promoted showed that lecturers did use technologies such as *YouTube* (56.7%), instant messaging (IM) apps such as WhatsApp (54.0%), discussion forums (37.8%), and video developmental tools (21.6%) to some extent (Vasodavan, DeWitt & Alias, 2019). However, tools such as blogs (Ifinedo, 2017; Shana & Abulibdehb, 2015) and podcasts for engaging and knowledge construction and co-production of learning materials (Bolliger, Supanakorn & Boggs, 2010; Middleton, 2016) were used less. Although these lecturers had the technology skills and knowledge to use technologies, they did not seem to have knowledge on pedagogies such as collaborative learning to implement it effectively (Vasodavan et al., 2019).

The emphasis in HEIs should be on acquiring skills for interactions, evaluation, and creation of new knowledge, problem-solving, and higher-level thinking (DeWitt, Siraj, & Alias, 2014; Ronen & Pasher, 2011). However, technologies employed for delivery of instruction seemed to focus more on the transmission of factual knowledge and lecturers did not seem to have the pedagogical knowledge and skills for teaching (Vasodavan et al., 2019). Lecturers are expected to have the pedagogies, and specifically, the technology pedagogical content knowledge (TPACK) in integrating technology into their lessons (Mishra & Koehler, 2006). However, many lecturers have not been trained for pedagogy and technology skills in designing CL environments (Bower, 2016; Gast, Schildkamp, & Van Der Veen, 2017; Hobbs & Tuzel, 2017).

In some universities, there have been some professional development for new lecturers in the HEI in the context to be trained in pedagogies and TPACK. However, training is normally limited. In the university in the context of the study, the training is completed in less than a week and there might be no opportunities for application of their training in real scenarios. Hence, lecturers may not have sufficient TPACK to implement online as well as blended learning effectively. This has also resulted in HEIs just implementing the online platforms preferred by their lecturers, without giving the learner much choice as they had to quickly adapt to it.

3.2.2 Emergency Remote Teaching

A pedagogy for online learning needs to ideally allow learners to go through phases of socialization and motivation before the sharing of content and development of new knowledge in an online community (Salmon, 2002). However, emergency remote teaching differs as it is a response to a crisis or disaster such as during the COVID-19 pandemic. Arising from the threat of COVID-19, many colleges and universities opted to move their courses online, and most of their faculty had to adapt and experiment new solutions to learning (Hodges, Moore, Lockee, Trust & Bond, March 27, 2020). However, emergency remote teaching needs to be differentiated from online learning or blended learning (Hodges et al., 2020).

According to Hodges et al. (2020), emergency remote teaching (ERT) is only a temporary situation with temporary instructional supports, when the instructional delivery needs to shift to an alternate delivery mode because of a crisis and would return to the normal face-to-face or blended learning format once the emergency is over. Hence, the purpose of ERT is not to design an online learning system but only to provide temporary access to instruction with instructional support that is set up quickly and made available during an emergency (Hodges et al., 2020) ERT has been noted in Afghanistan, when education which had been disrupted by wars could be delivered to children via radio education and through DVDs (Davies & Bentrovato, 2011). On the other hand, the design of a course or a program needs to be carefully planned and envisioned so that the delivery modes, methods, media, and resources meets the needs of the learners (Dick, Carey & Carey, 2015).

During the period when online learning became inevitable, many organizations put together their efforts and started developing guidelines for teaching online. The Association for Educational Communications and Technology (2020) provided resources for educators in response to COVID-19 pandemic. A prominent member, Jonathan Michael Spector (2020), also put forward 10 suggestions for educators new to online learning, among which were getting learners to know each other and their interests first and foremost. Davidson and Harris (Oct. 29, 2020) document that courses which focused on building learning communities in remote learning during the time of pandemic managed to engage learners in colleges and universities. Building an online learning community to support learners with social support is an important aspect of online learning (Hodges et al., 2020).

However, during the pandemic, professional development courses could not be the traditional courses that applied online models of instruction for online distance learning; what was needed was a rapid development of support and professional development to take into account large numbers of faculty in the institution (Hodges et al., 2020). Instructional support units had to urgently relook at how they worked and supported faculty during the crisis (Hodges et al., 2020). At the same time, they had to ensure that the institution was accountable to the learners for the achievement of the learning outcomes

Face-to-face learning, which had originally been supported in an ecosystem with library resources, social resources from peers, tutors, and instructors, among others, as well as a residential system for most of the learners in higher education would differ in an online learning situation (Hodges et al., 2020). The online distance learning system will not have a need for residential support, but instead other learner supports are required and investment is needed in such areas as the delivery of content and social support for the community of online learners. Most experts suggest a mix of synchronous sessions such as video conferencing and asynchronous sessions such as on discussion forums and note that online sessions should be interactive but avoid lecture-style communication approaches (Spector, 2020).

3.2.3 Transferability and Viability of Technology Use

A particular technology approach needs to be considered carefully, especially when it is implemented on a broad scale, such as the implementation of online learning for all HEIs during the pandemic. When considering the implementation of a technological solution, institutional managers need to consider the outcomes of the solution: Would the solution have similar outcomes as in a developed country when it is transferred to the developing country (DeWitt & Alias, 2019a). The *innate transferability of technological affordances* (Lubin, 2018) notes that difficulties may exist as the hardware, software, and infrastructure may not be compatible with that in the developed country. For example, video conferencing on Zoom and other platforms are commonly used in HEIs internationally. However, is this platform compatible to be used for delivery of instruction among learners in our institutions? One needs to consider where the learners are located during the pandemic. Many learners moved from residential colleges in their campus, commonly where undergraduates are housed, and went home, some of them to rural communities where electrical supply and internet connectivity is a problem. Videoconferencing, and even high-definition videos on *YouTube* may provide clarity in explanations, are interactive, and can motivate learners when delivered online. However, the broadband internet access may be a problem in some areas in developing countries like Malaysia. As such, there are anecdotes in this regard of learners who had to go to higher ground and climb on trees to obtain better signal for reception.

In cases like these, alternatives such as delivering applications or videos onto the computer or servers by installing them on the computer or laptop and delivering the materials on DVD and other devices can be a viable alternative (DeWitt & Alias, 2019b; Davies, & Bentrovato, 2011). Institutions, such as the one in the context of the study, made arrangements for learners who required access to be provided with internet connectivity through collaborations with Internet Service Providers, and for learners and lecturers who did not have functioning devices, to be provided with purchase schemes to assist in the procurement. Hence, consideration and adjustments would have to be made to ensure that the expected technologies which would be used for online learning would have a similar *innate transferability of technological affordances* when used during emergency remote teaching.

Another consideration Lubin (2018) puts forward is whether there is *viability of the technological innovation*. This would be related to the users' needs and the social effect on the users, in particular the learners and lecturers involved in the emergency remote teaching and learning. The social effect would be related to the acceptance level. When the HEIs went on the online teaching mode, was consideration made to determine the vision of how and what technologies were implemented (DeWitt & Alias, 2019a)? And during implementation, was the effect originally envisioned similar to what was expected? This is an important consideration as an instructional leader who inspires the institution in achieving a shared vision would have better opportunities for success (DeWitt & Alias, 2019b).

The social effect for the *viability* of the solution in this case, which is the online learning, is related to the people. The instructional leader not only inspires the institution with a shared vision but develops the vision by ensuring that the lecturers and learners have the right support, such as the related policies, to continue (DeWitt & Alias, 2019b). The institution in the context of the study already had existing information technology policies related to e-learning as well as privacy and security. However, policies related to the design of instruction were not clearly stated as the focus was on blended learning. On the other hand, several scenarios of disaster management and opportunities to rehearse for such disasters in the e-learning week conducted each semester provided lecturers the experience in handling a total online scenario. Hence, these existing policies could support the lecturers and learners when the learning shifted online.

Another area of viability is to develop instructors' capacities in technology and pedagogy. The actual test of the viability would be when the lecturers were able to use the technologies confidently, without many problems. This confidence would only develop when the instructors were competent and skilled. Hence, providing for professional development in technology and pedagogy is important (DeWitt & Alias, 2019b). Thus, during the period when learning shifted online, MOHE encouraged the HEIs to share webinars that were conducted for professional development so that the community of lecturers could share best practices in technology and pedagogy and improve their TPACK.

When the COVID-19 pandemic started, lecturers may have had difficulty in designing their online learning environments. This is evidenced in situations when instructional materials were provided without an applicable instructional model and a teacher-training program becomes less effective (DeWitt & Alias, 2019a). Hence, there were many webinars for professional development conducted for supporting the faculty. Perhaps it was not for the lack of online professional development but more on the variety of tools and approaches to use. There were different videoconferencing tools compared (*Microsoft Teams*, *Google Meet*, and *Zoom*) and regular webinars where the cost-effectiveness of the tools were highlighted and shared, and guidelines published. However, the face-to-face support was lacking, and many lecturers did what they could with whatever support that was in place.

Additionally, the availability of suitable resources for online teaching may also affect the viability of the innovation as materials might need to be adapted to the language of the developing country as well as the culture, morals, and customs (DeWitt & Alias, 2019a). During the pandemic, materials such as videos for the content may not be available in the medium of instruction, Bahasa Melayu, and this would affect the viability as the learners would need additional effort to understand the language and might lead to them not accepting the solution. There was a rush among lecturers to review and adapt their materials for an online learning environment. This led to additional workload among lecturers who were already burdened by the pandemic and the training.

Finally, in order to ensure the viability of the solution, the support and involvement of the community is needed (DeWitt & Alias, 2019a). In the HEI in the context of the study, members of the community, such as the lecturers, interested in learning the new technologies and pedagogies joined *WhatsApp* and *Telegram* groups where information, training, and other important concerns were broadcast to the community. This became a professional online Learning Community that provided some support and spaces for learning.

3.3 METHODOLOGY

This research utilized a cross-sectional survey method. A convenience sampling procedure was used for sample selection. The questionnaire used in this study was divided into two sections: (1) Demographic section focusing on respondents' gender, age, ethnicity, education level, and their location and (2) Section focusing internet

TABLE 3.1
Learners' Demographic Profile ($N = 251$)

Demographic		Frequency	Percentage (%)
Gender	Male	33	13
	Female	218	87
Age	Under 30 years	180	71
	30–39 years	42	17
	Above 39 years	29	12
Ethnicity	Malay	94	37
	Chinese	77	31
	Indian	43	17
	Bumiputera	10	6
	International student	22	9
Educational level	Undergraduate	78	31
	Postgraduate	173	69
Location	Urban	180	71
	Rural	50	19
	Remote	21	10

connectivity, computer skills, and learners' satisfaction with emergency remote teaching. All items in this section were close-ended items with a five-point Likert-type rating scale. The questionnaire was administered online and 251 respondents from a public university in Klang Valley, Malaysia, completed the survey. With regard to ethical consideration, learners' consent to participate in this study was sought on the front cover of the online questionnaire. Learners were given a choice either to take part in the survey or not. Participation was strictly voluntary and anonymous. Thus, by completing the questionnaire, the learners have given their consent. The demographic profile of the learners is shown in Table 3.1.

Next, data from the online questionnaire was exported into a Microsoft Excel file. Subsequently, the file was imported into WINSTEPS version 3.73, a Rasch measurement model software (Linacre, 2012). The raw data was then calibrated and validated using Rasch model analysis to ensure accurate and precise measurement (Bond & Fox, 2015). There was no missing value, outlier, and extreme misfit response in the data.

Rasch model transforms raw ordinal data (Likert-type data), based on frequency of response which appears as probability, to logit data (logarithm odd unit) via the logarithm function. Ordinal data contains scaling and nonlinearity issues that were addressed with the Rasch model linear measure (Andrich & Marais, 2019). Particularly, partial Credit Model (PCM), a type of Rasch model analysis was utilized to analyze two different types of rating scales in the instrument (Boone et al., 2014). The calibration and validation process of the data were done with PCM by examining the fit statistics of the items and persons, looking at indices infit and outfit mean squares (MNSQ), standard error measurement, residuals, rating scale analysis, and reliabilities of person and items (Boone et al., 2014; Sumintono & Widhiarso, 2015). For instance, for detecting the fit statistics of the items and persons, the index of infit and outfit of the MNSQ value should be 1.0; however, a range of 0.5–1.5 is still acceptable (Boone et al.,

2014). In addition, the data's standard error measurement indices were below 1.0 logit, indicating a goodness-of-fit to the model (Bond & Fox, 2015).

3.4 FINDINGS

The results of the items and person measure (logit) in each construct are presented in a Wright map. The Wright map essentially provides a linear ruler starting from zero logit to maximum logit where person location (LVP) is on the left and item position (LVI) is on the right (Englehard, 2013). For item calibration, the logit value of item (LVI) informs about the difficulty level. A higher LVI means that the item is difficult to reach agreement, while a low LVI means that the item is easy to reach an agreement. As for logit value person (LVP), a higher LVP indicates the person chose a higher score. Both the LVI and LVP use a linear measure in the same scale that can be presented in a Wright distribution map, which illustrates the item difficulty and person's agreement to each item. This reveals a deeper insight into the calibration of item and person (Boone et al., 2014; Sumintono & Widhiarso, 2014).

Figure 3.1 illustrates the Wright map in terms of internet connectivity and learner's computer skills. This Wright map consists of three items (i.e. IC, SC and SM). Item IC has the highest logit (LVI = 1.05 logit), indicating that many learners has difficulty in terms of internet connectivity. Seventy-one respondents had less than LVP=1.05 logit. In terms of learner's skill in using a computer (item SC) or a smartphone (item SM), results indicate that learners' skills in using a smartphone (LVI = −0.28 logit) are far better than that using a computer (LVI = 0.50 logit).

As for the most common videoconferencing tool used, most learners selected Google Meet (206 learners) followed by Microsoft Teams (139 learners) and Zoom (128 learners). In terms of learners' satisfaction with emergency remote teaching, the Wright map consists of five items (i.e. P1, P2, P3, P4, and P5). Item P3 has the lowest logit (LVI = −0.54 logit), indicating that majority of learners (95%) were able to receive lectures with an online learning platform. Only 13 learners (5%) have lower logit, indicating they are not able to receive lectures with an online learning platform. Next, item P1 (LVI = −0.31 logit) shows majority of learners (93%) are satisfied with the online learning platforms used, whereas only 19 learners (7%) were not satisfied with it.

Figure 3.2 shows the Wright map about students' satisfaction. As for item P2 and P5 (LVI = −0.21 logit), learners indicated they were able to communicate and find online learning platforms useful for learning. Only 19 learners (7%) have a low LVP score, indicating there are a few learners who struggle and find it difficult to learn with online learning platforms. The highest logit for this dimension was for item P4 (LVI = −0.01 logit), which queried learners to what extent they were able to share presentation and files online. Most learners rated all the platforms as satisfactory, though 28 learners (11%) responded they find it challenging to share presentations and files online.

3.4.1 Learners' Satisfaction with Emergency Remote Teaching

This section examines learner's satisfaction with emergency remote teaching. LVP was applied to obtain a mean and standard deviation of learner's satisfaction of emergency remote teaching. The mean of LVP (+2.47 logit) and standard deviation (2.15)

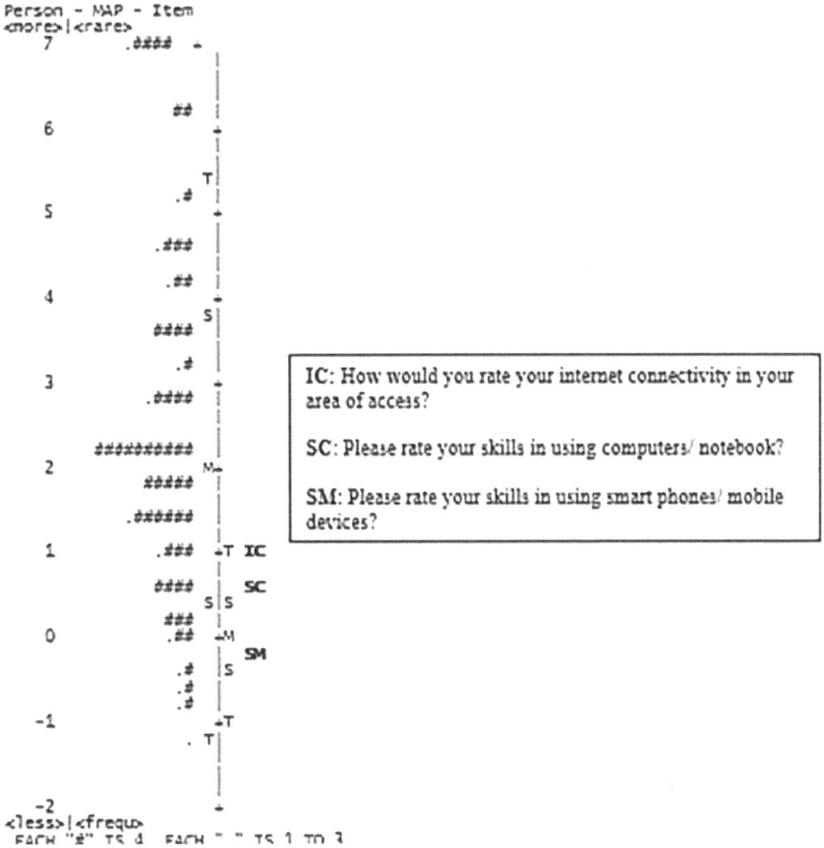

FIGURE 3.1 Wright map of learner's internet connection and computer skills.

was then used to classify learners' satisfaction level with emergency remote teaching (see Table 3.2). The satisfaction level was categorized into three groups, very satisfied, satisfied, and not satisfied, using the demographic profile of the respondents. Table 3.2 shows the details of learners' satisfaction level of emergency remote teaching with their demographic profiles.

Table 3.2 demonstrates that, regardless of demographic profile differences, most of the learners are very satisfied and satisfied with emergency remote teaching. However, some distinctive features were found in education level. Interestingly, postgraduate learners show a higher proportion of satisfaction as compared to undergraduate learners. Learners' gender, age, ethnicity, and location led to minimal differences in their satisfaction level.

3.5 DISCUSSION

The present study investigated learners' satisfaction, needs, and challenges with remote teaching, such as in regard to issues of connectivity, devices, and learning

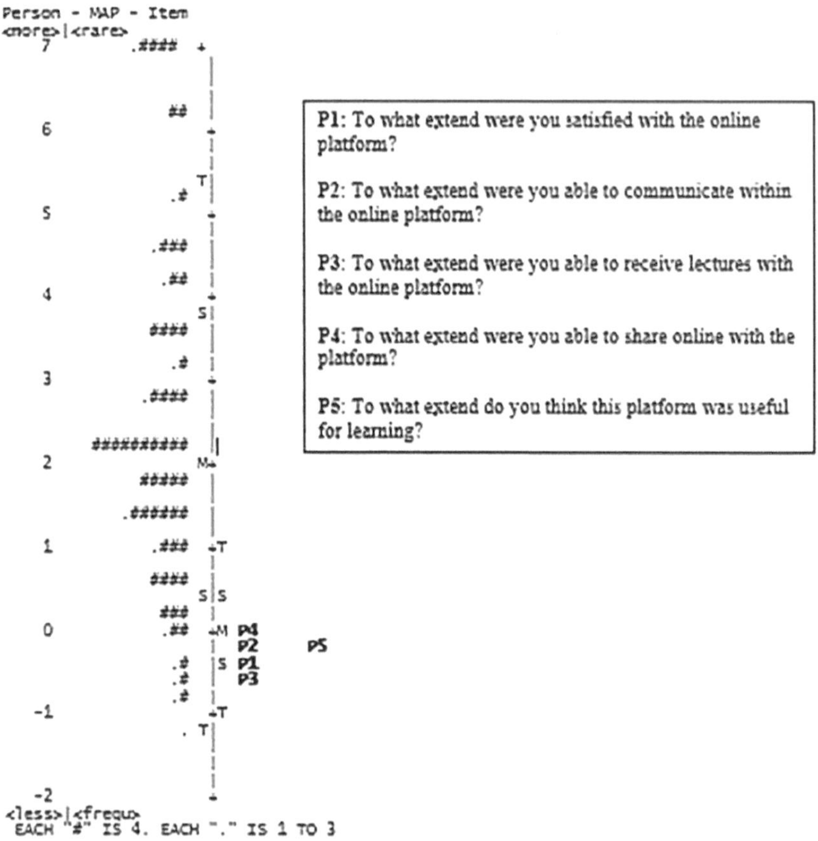

FIGURE 3.2 Wright map of learners' satisfaction with emergency remote teaching.

strategies, in a prominent public higher education institution during the emergency. Several important findings are discussed extensively in what follows.

First, just above a quarter of learners (28%) reported challenges in terms of low internet connectivity during emergency remote teaching. The findings are similar to those reported by leaners in private higher learning institutions in Malaysia (Yusuf & Jihan, 2020) and those in other developed and developing countries such as the Netherlands and Ghana who had limited access to the internet and adequate learning environments during this emergency period (Bol, 2020; Owusu-Fordjour et al., 2020). Hence, consideration and adjustments need to be made by HEIs to collaborate with Internet Service Providers to provide learners with stable internet connection. Alternatively, delivering instructions through videos on DVDs or other devices can be an alternative (DeWitt & Alias, 2019b; Davies & Bentrovato, 2011) and these courses needs to be carefully planned to ensure the delivery modes meet the needs of the learners (Dick, Carey & Carey, 2015).

Second, the result from this present study show that learners are more skilled in using their smartphone for learning as compared to computers. This is hardly surprising as most learners in this study are from a technology-savvy generation. This

TABLE 3.2
Learners' Satisfaction of Emergency Remote Teaching (*N* = 251)

Demographic	Very Satisfied LVP > +2.47	Satisfied +2.47 > LVP > +0.32	Not Satisfied LVP <+0.32
Gender			
Male	15	11	7
Female	80	105	33
Age			
Under 30 years	69	89	22
30–39 years	15	18	9
Above 39 years	11	9	9
Ethnicity			
Malay	30	51	13
Chinese	38	28	11
Indian	13	21	9
Bumiputera	2	7	2
International student	9	8	5
Education level			
Undergraduate	20	48	10
Postgraduate	75	68	30
Location			
Urban	75	78	27
Rural	15	28	7
Remote	5	10	6

is consistent with the findings of Adams et al. (2020a) who showed that smartphone was the preferred device utilized for learning by learners. However, this requires further exploration as the majority of the learners who participated in the present study are postgraduate learners; thus, discrepancies exist when compared with Adams et al.'s (2018) study which found that undergraduate learners prefer smartphones for learning slightly more than postgraduate learners.

Next, we found *Google* Meet as the most common videoconferencing tool used by learners. This videoconferencing tool seems a popular choice among most learners. They are very satisfied with this tool as they are able to attend, learn, and communicate effectively in online classes and lectures. However, a few learners were dissatisfied with *Google Meet* as they found it difficult to present and share files online. This might have been because these learners were more familiar with the features and navigation in other videoconferencing tools they had started with, rather than in *Google Meet*. In Permata and Bhakti's (2020) study, *Google Meet* was also reported as an effective videoconferencing tool for remote emergency teaching and learning. Suhada et al. (2020) support these findings with *Google Meet* being rated as an excellent and effective tool for teaching during the COVID-19 pandemic.

Fourth, this study found that most of the learners were satisfied with remote teaching. Learners' demographic characteristic such as gender, age, ethnicity, and location had only minimal impact on their level of satisfaction with remote teaching. This finding could be due to the existing infrastructure and support at the HEI in the context of this study that was stable and able to support the learners in emergency remote teaching mode. Access to *Microsoft Teams* and *Google Meet* were available to the community of lecturers and learners in addition to the Learning Management System. Hence, the technological infrastructure seemed to be in place for the online learning.

Additionally, a high proportion of postgraduate learners in this study preferred remote teaching style as compared to undergraduate learners. Similar findings were also found by Adams et al. (2018; 2020b) in which mature students were found to prefer online learning as an alternative to the typical face-to-face interaction, read materials online, and prepare for classes in this manner. It can be interpreted that postgraduate learners are more mature and independent learners as compared to undergraduate learners.

3.6 CONCLUSION

In conclusion, this study contributes to the existing literature on emergency remote teaching by providing empirical evidence leading to new knowledge in the Malaysian context. Consideration and adjustments need to be made by HEIs to provide stable internet connectivity for all learners. Particularly, further support and encouragement should be extended to undergraduate learners to help them in the transition to emergency remote teaching. This study has important implications for lecturers and HEIs. Lecturers and course administrators must consider learners who are adapting to this change themselves and may not be able to attend online classes immediately or regularly. Instead of synchronous activities, asynchronous activities should be prioritized (Hodges et al., 2020). Flexibility to assignment deadlines, adjustments to learners' learning outcomes, and alternative assessments should be accorded.

Future safety and health concerns such as natural disasters and pandemics may happen. Thus, continuous professional development courses for emergency remote teaching must be continued to prepare university lecturers and HEIs for campus closure. Emergency remote teaching must become part of a lecturer's skill set and be included in the instructional design of universities' courses (Hodges et al., 2020). Lecturers and administrators need to consider the *viability of the technology innovation* when considering technology tools for learning so that the *innate transferability of technological affordances* of these tools could be applied to different situations (Lubin, 2018).

This study has several limitations. First, data was collected from only one public university in Klang Valley, Malaysia. Such limitation prohibits generalization of the findings to a broader context. Future studies could include more representative data to ensure generalization of the findings by collecting data across 20 public universities in Malaysia using probability random sampling. Second, the current study is limited to only a quantitative cross-sectional research design. As such, future studies could adopt a mixed-methods research design to further explore and validate the quantitative data findings. In essence, this study's empirical evidence could be used as future reference for emergency remote teaching either in the Malaysian or international context.

REFERENCES

Adams, D., & Dewitt, D. (2021). *Innovative Practices of Technology-Enhanced Learning.* Universiti Pendidikan Sultan Idris.

Adams, D., Tan, M. H. J., & Sumintono, B. (2020a). Students' readiness for blended learning in a leading Malaysian private higher education institution. *Interactive Technology and Smart Education.* https://doi.org/10.1108/ITSE-03-2020-0032.

Adams, D., Sumintono, B., Mohamed, A., & Noor, N. S. M. (2018). E-learning readiness among students of diverse backgrounds in a leading Malaysian higher education institution. *Malaysian Journal of Learning and Instruction, 15*(2), 227–256.

Adams, D., Tan, M. H. J., Sumintono, B., & Oh, S. P. (2020b). Blended learning engagement in higher education institutions: A differential item functioning analysis of students'backgrounds. *Malaysian Journal of Learning and Instruction, 17*(1), 133–158.

Andrich, D. and Marais, I. (2019). *A Course in Rasch Measurement Theory, Measuring in the Educational, Social and Health Sciences.* Dordrecth: Springer.

Asad, M. M., Hussain, N., Wadho, M., Khand, Z. H., & Churi, P. P. (2020). Integration of e-learning technologies for interactive teaching and learning process: an empirical study on higher education institutes of Pakistan. *Journal of Applied Research in Higher Education.* https://doi.org/10.1108/ JARHE-04-2020-0103.

Association for Educational Communications and Technology (2020). AECT's Response to the COVID-19 Virus. https://www.aect.org/aects_response_to_the_covid-1.php.

Biasutti, M. (2017). A comparative analysis of forums and wikis as tools for online collaborative learning. *Computers & Education, 111*, 158–171.

Bol, T. (2020). Inequality in homeschooling during the Corona crisis in the Netherlands. First results from the LISS Panel.

Bolliger, D., Supanakorn, S., & Boggs, C. (2010). Impact of podcasting on student motivation in the online learning environment. *Computers and Education, 55*(2), 714–722. http://doi.org/10.1016/j.compedu.2010.03.004.

Bond, T. G., & Fox, C. M. (2015). *Applying the Rasch Model, Fundamentals Measurement in the Human Sciences* (3rd ed.). New York: Routledge.

Boone, W. J., Staver, J. R., & Yale, M.S. (2014). *Rasch Analysis in the Human Sciences.* Dordrecth: Springer.

Bower, M. (2016). Deriving a typology of Web 2.0 learning technologies. *British Journal of Educational Technology, 47*(4), 763–777. https://doi.org/10.1111/bjet.12344.

Burhan-Horasanlı, E., & Ortaçtepe, D. (2016). Reflective practice-oriented online discussions: A study on EFL teachers' reflection-on, in and for-action. *Teaching and Teacher Education, 59*, 372–382.

Davidson, C. N. & Harris, D. (Oct 29, 2020). Making Remote Learning Relevant. Inside Higher Ed. Retrieved from https://www.insidehighered.com/views/2020/08/05/colleges-should-throw-out-conventional-pedagogical-and-curricular-playbook-fall.

Davies, L. & Bentrovato, D. (2011). Understanding education's role in fragility: synthesis of four situational analyses of education and fragility: Afghanistan, Bosnia and Herzegovina, Cambodia, Liberia. Paris: UNESCO- International Institute for Educational Planning.

DeWitt, D., & Alias, N. (2019a) Computers in education in developing countries: Managerial issues. In Tatnall, A. (Eds.). *Encyclopedia of Education and Information Technologies.* Springer, Cham

DeWitt, D., & Alias, N. (2019b) Support for school and institutional improvement and accountability. In Tatnall, A. (Eds.). *Encyclopedia of Education and Information Technologies*. Cham: Springer.

DeWitt, D., Alias, N. and Siraj, S. (2015), Collaborative learning: Interactive debates using padlet in a higher education institution", In *Proceedings of the International Educational Technology Conference*, pp. 88–95, Istanbul, Turkey.

DeWitt, D. & Koh, E.H.Y. (2020), Promoting knowledge management processes through an interactive virtual wall in a postgraduate business finance course. *Journal of Education in Business*, 95 (4), 255–262.

DeWitt, D., Siraj, S., & Alias, N. (2014). Collaborative mlearning: A module for learning secondary school science. *Educational Technology and Society*, *17*(1), 89–101.

DeWitt, D., Alias, N., Siraj, S., & Hutagalung, F. (2014). Knowledge Management in Education : Designing instruction for using wikis. In: Gaol, F.L., Mars, W. & Saragih, H. (Eds.), *International Conference on Management and Technology in Knowledge, Service, Tourism & Hospitality* (pp. 45–50). London: Taylor & Francis Group.

Dick, W., Carey, L., & Carey, J. O. (2015). *The Systematic Design of Instruction* (8th Ed.). New York: Pearson.

Englehard, G. (2013). *Invariant Measurement, Using Rasch Models in the Social, Behavioral, and Health Sciences*. New York: Routledge.

Gast, I., Schildkamp, K., & Van Der Veen, J. T. (2017). Team-based professional development interventions in higher education: A systematic review. *Review of Educational Research*, 87(4), 736–767. https://doi.org/10.3102/0034654317704306.

Hobbs, R., & Tuzel, S. (2017). Teacher motivations for digital and media literacy: An examination of Turkish educators. *British Journal of Educational Technology*, *48*(1), 7–22. https://doi.org/10.1111/bjet.12326.

Hodges, C., Moore, S., Lockee, B., Trust, T., & Bond, A. (2020, March 27). The Difference Between Emergency Remote Teaching and Online Learning. Educause Review. Retrieved from https://er.educause.edu/articles/2020/3/the-difference-between-emergency-remote-teaching-and-online-learning

Ifinedo, P. (2017). Examining students' intention to continue using blogs for learning: Perspectives from technology acceptance, motivational, and social-cognitive frameworks. *Computers in Human Behavior*, 72, 189–199. http://doi.org/10.1016/j.chb.2016.12.049.

Middleton, A. (2016). Reconsidering the role of recorded audio as a rich, flexible and engaging learning space. *Research in Learning Technology*, *24*(1063519), 1–14. http://doi.org/10.3402/rlt.v24.28035.

Noor, A. M., Attaran, M., & Alias, N. (2015). Students' experiences in using spectrum: Textbook or classroom? *Procedia – Social and Behavioral Sciences*, *176*, 667–673. doi: 10.1016/j.sbspro.2015.01.525

Li, C., & Lalani, F. (2020). The COVID-19 Pandemic Has Changed Education Forever. This Is How. World Economic Forum. https://www.weforum.org/agenda/2020/04/coronavirus-education-global-covid19-online-digital-learning/.

Linacre, J. M. (2012). *A User's Guide to Winsteps Ministeps Rasch-Model Computer Programs [version 3.74.0]*. Chicago, IL: Winstep.com.

Lubin, I. A. (2018). Global forces, local needs, and ICT supported innovations in small countries and developing regions: Overcoming the misalignment. An editorial introduction. In Lubin, I. A. (Ed.). *ICT-Supported Innovations in Small Countries and Developing Regions: Perspectives and Recommendations for International Education* (pp. 3–26) Cham, Switzerland: Springer International.

Ministry of Higher Education (MOHE). (2015). *Malaysia Education Blueprint 2015–2025 (Higher Education)*. Putrajaya: Ministry of Education Malaysia. https://doi.org/10.5923/j.ijis.20120206.05.

Mishra, P., & Koehler, M. J. (2006). Technological pedagogical content knowledge: A new framework for teacher knowledge. *Teachers College Record*, *108*(6), 1017–1054.

Owusu-Fordjour, C., Koomson, C. K., & Hanson, D. (2020). The impact of Covid-19 on learning-the perspective of the Ghanaian student. *European Journal of Education Studies*, 7, 88–101.

Permata, A., & Bhakti, Y. B. (2020). Keefektifan virtual class dengan google classroom dalam pembelajaran fisika dimasa pandemi Covid-19. *JIPFRI (Jurnal Inovasi Pendidikan Fisika Dan Riset Ilmiah)*, *4*(1), 27–33.

Ronen, T., & Pasher, E. (2011). *Complete Guide to Knowledge Management a Strategic Plan to Leverage your Company's Intellectual Capital*. Hoboken, NJ: Wiley.

Salmon, G. (2002). Mirror, mirror, on my screen≡ Exploring online reflections. *British Journal of Educational Technology*, *33*(4), 379–391.

Shana, Z. A., & Abulibdehb, E. S. (2015). Engaging students through blogs: Using blogs to boost a course experience. *International Journal of Emerging Technologies in Learning*, *10*(1), 30–38. http://doi.org/10.3991/ijet.v10i1.4240.

Spector, J. M. (2020). Advice for Those New to Online Learning. (https://aect.org)

Suhada, I., Kurniati, T., Pramadi, A., & Listiawati, M. (2020). Pembelajaran daring berbasis Google Classroom Mahasiswa Pendidikan Biologi pada Masa wabah Covid-19. *Digital Library UIN Sunan Gunung Djati*, 1–10.

Sumintono, B. & Widhiarso, W. (2014). *Aplikasi Model Rasch untuk Penelitian Ilmu-Ilmu Sosial (edisi revisi) [Application of rasch modelling in social science research, revised edition]*. Cimahi: Trimkom Publishing House.

Sumintono, B. & Widhiarso, W. (2015). *Aplikasi Pemodelan Rasch pada Assessment Pendidikan [Application of rasch modelling in educational measurement]*. Cimahi: Trimkom Publishing House.

UNESCO (2020). 1.3 Billion Learners Are Still Affected by School or University Closures, As Educational Institutions Start Reopening Around the World, Says UNESCO. https://en.unesco.org/news/13-billion-learners-are-still-affected-school-universityclosures-educational-institutions

Vasodavan, V., DeWitt, D., & Alias, N. (2019). TPACK in higher education: analysis of the collaborative tools used by lecturers. *Jurnal Kurikulum & Pengajaran Asia Pasifik*, *7*(1), 9–17.

Yusuf, B. N., & Jihan, A. (2020). Are we prepared enough? A case study of challenges in online learning in a private higher learning institution during the Covid-19 outbreaks. *Advances in Social Sciences Research Journal*, *7*(5), 205–212.

4 Advanced Instructional Strategies in a Virtual Chemistry Laboratory during the COVID-19 Outbreak

Payal B. Joshi
SVKM's NMIMS University

CONTENTS

4.1 INTRODUCTION

The severe contagious nature of the novel coronavirus and announcement of lockdown due to the COVID-19 pandemic led to closure of higher educational institutes, universities, and colleges so as to follow the guidelines of social distancing (Anderson et al. 2020; Murphy 2020). This resulted in an impediment to the in-campus teaching and laboratory practices, which affected over 900 million students globally (UNESCO 2020). Academic leadership and policymakers involved in the education sector suddenly were compelled to ensure remote learning and assessments continued for the students with utmost urgency. Teachers and educators were challenged to continue learning and assessment of the students. It was witnessed that the transition from conventional to virtual teaching was far from being smooth. In fact, laboratory-based disciplines suffered the most as major experimental sessions

DOI: 10.1201/9781003143796-4

got suspended for an uncertain period of time. Among science subjects, chemistry, physics, and biochemistry faced crucial issues of conducting online laboratories for the students. Chemistry is a highly complex and conceptual subject. A chemistry learner needs to acquire understanding at observational, experiential, instrumental, and symbolic levels. Chemistry learners need to develop advanced spatial abilities and visual literacy (Harle and Towns 2011). They need to acquire "hands-on" experience such as performing titrations, synthesizing organic compounds, and analyzing mixtures using advanced instrumentation techniques. The transition of chemistry laboratories to an online mode was not a straightforward endeavor. In India, universities have unique chemistry curricula that made it difficult to choose a typical standalone virtual platform encompassing all the experiments. Thus, the choice of a typical chemistry experiment on a virtual platform became a challenging task for chemistry educators. The pandemic situation seemed to be lurking for an uncertain timeline that left no choice but to continue laboratories through a virtual mode entirely.

The present chapter discusses about virtual chemistry laboratories and advanced pedagogical strategies while conducting virtual screen experiments particularly for engineering students on organic synthesis, stereochemistry, and spectroscopy. A detailed account on assessment strategies during virtual experiments is presented. It concludes with challenges, opportunities, and future trends of virtual chemistry laboratories.

4.2 SIMULATING IN A CHEMICAL ENVIRONMENT

A virtual laboratory is a computer-based program that allows interaction of a user with an experiment in a remote manner and not physical reality. Virtual laboratories gained importance as it is perceived as an inquiry-based and student-centric in design promoting advanced level of cognition and thinking. The major attraction of virtual laboratories is the low-cost alternative compared to a typical chemistry lab that requires expensive instruments and skilled manpower. Virtual labs allow recreating complex, expensive, and unsafe experiments that are usually not included in traditional chemistry experiments due to safety concerns (Achuthan and Murali 2015; Jones 2018; Tatli and Ayas 2010). Laboratory study is considered as an experiential learning as it dates back to Piaget's cognitive development theories (Piaget 1964) and Kolb's experiential learning strategies (Kolb, Boyatzis and Mainemelis 2001). Chemistry is an experimental science and involves many abstract concepts that necessitates laboratory-based evidences to understand and comprehend them in a clear way. This is the reason laboratory-based activities form an integral part of chemistry curriculum and, thus, chemistry experiments are designed in "learn by doing" strategy. The advanced-level chemistry graduates need to carry out experiments in a chemistry laboratory to "see" the concepts come in play. Until now, chemistry educators thought of replacing traditional chemistry laboratories with a simulated environment only if the requirements are costly. It was argued that students cannot learn experiments by merely clicking on-screen beakers and burettes. In developing countries like India, teaching science and advanced chemistry requires huge investments by universities, such as in terms of chemicals, laboratories equipped

TABLE 4.1
Selected Virtual Labs and Interactive Simulations in Chemical Sciences

Virtual Labs and Interactive Simulations	Features
MHRD Virtual labs, Government of India	Online experiments under physical, chemical sciences and other engineering disciplines.
PhET Interactive Simulations, University of Colorado	Simulation experiments for general and quantum chemistry.
Screen Experiments, Royal Society of Chemistry	Interactive simulation experiments on titrations and aspirin synthesis.
ChemCollective, Carnegie Mellon University	Scenario-based lab activities for stoichiometry, physical chemistry, thermochemistry, and chemical kinetics.
AACT Simulations, American Association of Chemistry Teachers	Multimedia for chemistry experiments and demonstration activities.
Molecular Workbench, The Concord Consortium	Advanced simulation labs for chemistry, physics, biotechnology, and nanotechnology that also allows creating virtual labs to educators.
MERLOT Materials: Chemistry Simulations	Interactive, research-based simulation labs covering large number of chemistry experiments.
CK–12 Chemistry Simulations	Simulation chemistry labs related to atomic masses, solubility studies, and demonstrations
Olabs, Ministry of Electronics and Information Technology, Government of India	Interactive simulation labs for science and mathematics.
Interactive 3D Chemistry Animations, University of Liverpool	Advanced animations for topics such as hybridization, spectroscopy, and organic reactions.
Mechanism Inspector, Royal Society of Chemistry	Simulations on organic reaction mechanisms to assess core investigative skills of the learner.

with safety tools, and analytical instruments. Virtual chemistry laboratories obviate this requirement to conduct experiments with costly instruments and hazardous chemicals. The major challenge is to simulate a typical chemistry laboratory so that the learner can "feel, see and experience," in a similar way as if the experiment was performed in a traditional laboratory. Table 4.1 lists some interactive simulations that are utilized for conducting chemistry virtual experiments.

The concept of virtual laboratory in chemistry is not new, though at quite a nascent stage. Recent instances revealed the use of virtual reality in chemistry education (Dunnagan, et al. 2020; Plunkett 2019). Until now, these online simulations served as "problem-based learning" or were incorporated in distance education to enhance student learning. However, there are numerous studies that indicate no significant differences in learnings between virtual and conventional laboratories (Jong, Linn and Zacharia 2013; Josephsen and Kristensen 2006). The current pivot toward virtual laboratory in pandemic times has unintentionally given rise to eliminating overall "hands-on" experience. Usually, chemistry simulations were used as a 'prelab" prior to conducting an actual experiment in a conventional setting. It was observed that sole access to online simulations did not serve the purpose of an actual laboratory experience to the students. It was essential to incorporate a robust teaching style while conducting online laboratories for the students along with precise grading policies. To put it simply, virtual laboratories are available all the time mimicking the actual experiments, but teachers have to guide, assess, and exercise the learning of the students. In the present situation, graduate chemistry labs are of a two-to-three-hour session which is necessary to be conducted online wholly. Thus, it is essential to have a clear, concise pedagogical design for virtual chemistry labs that pave the way for robust grading policies.

4.3 PEDAGOGICAL DESIGN FOR VIRTUAL CHEMISTRY LABORATORY

Apart from teaching theory, chemistry laboratory is an essential and crucial component in chemistry education. Several published reports revealed the need to incorporate experiments in chemistry and chemical sciences (Galloway and Bretz 2015; Hofstein, Nahum and Shore 2001; Nakhleh, Polles and Malina 2002; Reid and Shah 2007). Ideally, the curriculum of chemistry courses at the university level focuses on student outcomes, with little emphasis on fulfilling student expectations and faculty perspectives. An ideal chemistry education framework should possess all elements such as teacher, learner, content, context, and evaluation in unison. Several inconsistencies among teacher's perspectives, student expectations, and outcomes act as a major hindrance in active learning experiences in classroom and laboratories. Various factors impact student learnings such as student emotions, attitude toward the subject, and cognitive and psychomotor verbal skills (Brandriet, Ward and Bretz 2013; DeKorver and Towns 2015; Flaherty, et al. 2017).

In the present study, it was revealed that organic synthesis, stereochemistry, and spectroscopy are complex topics in chemistry for engineering students. The stages in a typical conventional lab setting are summarized as follows:

1. Arranging the chemicals, glassware, instruments, and apparatus for the experiment.
2. Providing safety data sheet to students prior to handling toxic chemicals and the precautions in case of an emergency such as spills or leakage.
3. Conducting laboratory activities and using chemicals and reagents that are monitored by the teacher.
4. Collecting data and processing of results by the students that are graded by the teacher.

Even with the above pointers, insufficient laboratory conditions limit the teachers' ability to conduct them often coupled with errors in data collection. As seen in Table 4.1, students were provided simulated experiments for performing chemistry experiments in the 15-week semester. Aspirin synthesis, polarimetric study of sugars, and ultraviolet-visible spectrophotometry are the three crucial experiments that fall under the category of organic synthesis, stereochemistry, and spectroscopy respectively. These experiments were considered challenging by the students; thus, its pedagogical methods are described.

4.3.1 Aspirin Synthesis

In this experiment, the objective was to simulate the preparation of aspirin by esterification reaction. The procedure for the experiment was provided as conventional lab and simulation laboratory styles. It was observed that simulation procedure dealt with the usage of phrases such as "click" or "drag", which was assisting the students to navigate onscreen simulation. The experiment involved heating the reaction mixture for 15 minutes, which was time-lapsed in the simulation. The crude aspirin product was recrystallized in a solvent to be chosen from a range of solvent choices that were provided in the simulation lab. Overall, if the teacher's perspectives are considered, "sense of time" was crucial to understand the nuances of the trial-and-error in organic compound preparations. In a traditional aspirin synthesis, reaction completions are monitored using thin-layer chromatography, which was provided in the simulation activity. Thus, providing nuances of reaction completion steps in a chemistry simulated environment are crucial features for a robust simulated laboratory experience.

4.3.2 Polarimetric Analyses of Sugars

The objective of the experiment was to determine optical activity and specific rotation of sugars using polarimetry. Very few simulation resources could suffice for the implementation of polarimetric analyses of sugars that were in accordance with our course. Hence, it was decided to record lab activity and provide it as a prerecorded demonstration video to the students. The students were shown in the video about the differences in optical activities of sugars by analyzing glucose and fructose solutions (Figures 4.1a and 4.2b, respectively) in a polarimeter.

Students were oriented with the use of Excel a week before conducting the polarimetric experiment. During the postlab activity, they were instructed to plot a graph of angle of rotation (α) against concentration of the optically active substance. Using a graphical method, students obtained slope and reported unknown concentration and specific rotation of sugar solutions accurately.

(a) Polarimetric reading for 10% (w/v) glucose solution

(b) Polarimetric reading for 10% (w/v) fructose solution

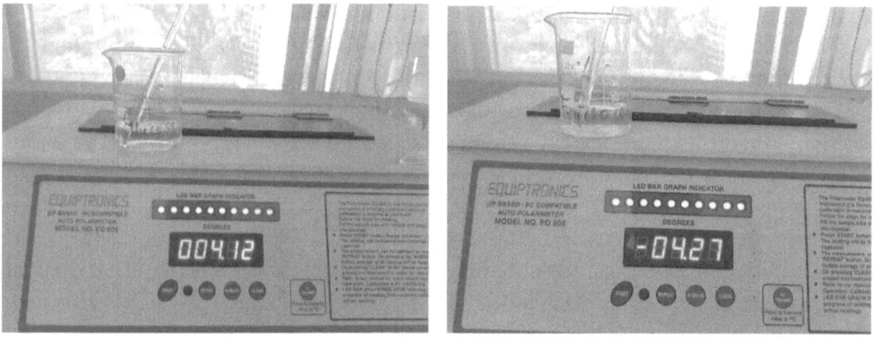

FIGURE 4.1 Polarimeter Analysis.

4.3.3 Ultraviolet-Visible (UV-VIS) Spectroscopy

The objective of the experiment was to determine absorbance maxima for various compounds using a simulated UV-VIS spectrophotometer. In this virtual activity, students recorded spectra of the four chemical compounds namely, cobalt chloride, ferrocene, crystal violet, and coumarin. It featured calibrating the spectrophotometer and verifying Beer-Lambert's law for the students in a virtual mode entirely. Further, students collected and analyzed the data from Beer-Lambert's plot on Excel along with reporting the unknown sample concentrations accurately. Students were also provided with unique UV-VIS spectra of the four compounds that provided insights about conjugation effects, characteristic colors, and solvent effects. Post all simulation activities, it was observed that simulations and prerecorded demonstrations had no marked influence on the students' learning and almost all teaching methodologies seemed satisfactory.

An online survey-based study of 43 engineering students of the first-semester chemistry course was conducted from 7 to 14 December 2020 to collect the information. A questionnaire link using "Google Forms" was sent to the students by email (Appendix 1). All the students provided full consent before participation and completed the survey with details of their experiences and suggestions. All participants had attended 2-hour chemistry laboratory sessions each week in a 15-week long semester (August–November 2020). A total of 12 laboratory sessions were conducted during the course, and the student batch size was kept at 10–12 students. It was revealed that about 69.8% of respondents found the screen experiments interesting, useful, and interactive as it gave error messages while performing certain steps but were navigated to the next level without teachers' assistance. Hence, in chemistry simulations, it is essential to allow learners to commit errors but not make them restart the level all over again. Further, about 62.8% of students found procedural complexity while performing aspirin synthesis as they needed clearer instructions for each mechanistic step. Hence, the assumption that chemistry simulations are easy and can be dealt with by students alone was a misconception found in the survey. On taking account of the faculty perspective, students need a "sense of time" as practically aspirin preparation usually takes 1 hour or more to complete steps as against the 30-minute simulation activity. However, it is futile to emulate long hours of chemical reactions on a simulator as the major focus. Thus, there is a need to orient students towards trial-error in organic synthesis and enhance their understanding by incorporating details of the reaction mechanism as an instruction mode. About 31.4% of students realized the lack of laboratory resources for performing the experiment, as they reported not "feeling" or "seeing" the product, especially in aspirin synthesis.

It was realized that to thoroughly understand the differences in a solitary simulation activity and teacher-blended instructed simulation, the study involved a control group ($n=20$) of students who were not oriented with teaching features such as using Excel for graphs, calculations, and reaction mechanism for these three experiments. An experimental student group ($n=23$) were oriented with these features prior to the simulation activity. We found that the control group were slower in reporting the graphical plots, calculations, reaction mechanism and spectroscopic data. This allowed for peer assessment in the spectroscopy virtual lab where the experimental group were assigned the role to monitor, navigate, and assess the control group toward faster and seamless simulation using graphical plots. Post-simulation activity, all students presented their findings as a laboratory report that was assessed and graded by the teacher. Later, students in the control group were oriented with all the teaching features to eliminate any bias in the study. Finally, both the student groups were assessed on a context-based test (or postlab quiz) that allowed us to understand the outcome in terms of experimental skills and analytical knowledge. Based on the survey outcomes, the pedagogical design proposed for conducting a seamless virtual chemistry lab are as follows:

a. Informing the students about the simulation website to be utilized for the experiment.
   Students should be aware of the simulated chemistry experiments in the chemistry course. Teachers must provide PreReads that serve as a background of their laboratory course.
b. Informing the system requirements necessary like flash plugins, extension add-ons, browser compatibilities, and related internet diagnostics.
   Just like in a conventional laboratory, glassware and apparatus requirements are crucial. A simulation chemistry laboratory should provide information about the technical details to complete lab work seamlessly.
c. Providing a detailed outline of the simulation procedure with mechanistic details of the reaction along with a format for reporting the experimental results.
   Teachers should design the simulation procedure for every experimental step. The simulation activity should allow students to identify errors especially in calculations and apply necessary corrective actions. If the simulation activity provides some attempts to complete a laboratory step, students can navigate with caution and avoid repetitive errors, if any.
d. Conducting a prelab quiz based on the experiment.
   A prelab quiz should be concise, clear, and well-framed as a short 5-minute activity. Teachers can design questions based on the complexity of the experiment. A short video presentation should be provided to the students about the activity. If graphs are required, teachers should orient the students to the features of Excel before the laboratory activity.
e. Performing an actual simulation by the students with virtual teacher assistance
   In a simulation, "Help menu," or "Avatar" should be provided in every step of the activity that creates a robust simulated environment.
f. Data collection and mechanistic details to be provided by the students.
   Teachers should provide the equations, chemical reactions, stoichiometric details, and physical data to allow students to derive the final results.
g. Final grading for the student lab report by the teacher.
   The simulation should allow maintaining lab records after completing experiments virtually. Students can download the completed lab record and report to the teacher for final assessment and grading.

The author acknowledges the use of smaller sample size as the study relied on students enrolled in the course. Also, using more sets of control and experimental student groups for all experiments is suggested in the future study so as to generalize the findings for improving the pedagogical design.

4.4 ASSESSMENT STRATEGIES FOR VIRTUAL CHEMISTRY LABORATORY

It has been observed for a long time that the traditional assessment strategies are criticized as they limit students to merely assimilating knowledge for obtaining grades with little room for developing skills and proficiency. At the time when COVID-19 in-campus restrictions were enforced, conducting final examinations became a major challenge for most universities (Roy and Roy 2020; Gamage, de Silva and Gunawardhana 2020). Online assessment and examinations were the only options either by proctored or an unproctored route. The current emergency situation allowed educators and policymakers to improvise online assessment strategies in laboratory sessions. Chemistry instructors devised assessment modes for lab sessions employing various rubrics for calculating student outcomes. Apart from fair grading policies for virtual simulations, there were serious concerns pertaining to academic schedules. Most colleges and universities suggested that rigorous academic schedule similar to those of traditional courses be followed stringently. This resulted in a negative impact on the teachers who were trying to cope with online teaching that was fairly new to them. Policy administrators, college authorities, and institutes seemed to throttle the adherence to a rigid academic calendar, proctored tests, and increased assignments, which worsened the problem. It is essential for the policymakers to understand that teachers need to deliver subject content and assess students in a rigorous manner, but not such that it burdens the students. The constant burdening of assignments and tests seem to diminish the students' enthusiasm for online learning. Many studies (Bao 2020, Rosário, et al. 2015) revealed that when subject content was fragmented and delivered by the teacher, students performed better in tests and other related laboratory activities. Thus, flexible timelines for lab report submissions, completion of quizzes, and flexible attitudes of teachers will result in better student engagement in the subject. Often, it is essential to provide autonomy to the teachers to decide their working style, and policies should evolve around them. Other alternative forms of assessments must be explored and incorporated that fosters positive student learning. In the face of crisis, it becomes extremely critical to maintain a student-centered perspective along with faculty perspectives while devising online assessment strategies. If assessment strategies place emphasis only on grades, it seems a myopic perspective; our main emphasis should on be the learning of students. It was also suggested that awarding letter grades to credit or no-credit provided flexibility to the students especially during the pandemic crisis (Sutton 2020). In the present study, virtually enhanced animations, interactive 3D models, and podcasts were blended in some laboratory sessions for assessing the learning outcomes of students. When these assessment tools were applied in a chemistry laboratory, we observed the students' relief as they had a break from their monotonous exam routine. This also provided for maintaining a fair grading policy for every outcome during simulated lab sessions and served as a typical transformed pedagogical assessment design. A prelab quiz was of concise nature with straightforward questions so as to build the confidence of the students. During simulation activities, students were oriented to the visual representation of chemical structures, reaction mechanisms, and instrumentation. A postlab quiz was designed to assess the understanding level of students and graded for every lab activity. The students received instant feedback on their simulated lab performance that gave an insight into their errors and chances of improvement in the next lab activity.

4.5 CHALLENGES, OPPORTUNITIES, AND FUTURE TRENDS

There are various challenges concerning online teaching such as lack of internet access, inconsistent power supply, internet speed, digital competency, lack of interactions (Adedoyin and Soykan 2020), and online security threats (Nam 2019). In India, with a population of 1.3 billion, only 600 million internet users are present, reflecting a major hindrance in providing online education especially in rural parts of the country (KANTAR IMRB 2019). The Indian government came up with robust measures to support educators in the country that allowed seamless virtual teaching. Under the aegis of the Ministry of Education, Virtual Labs allowed simulation activities for physical and chemical sciences. Yet, chemistry educators found many important chemistry laboratories missing. In such cases, teachers prerecorded their lab activities and uploaded them on open resources such as YouTube or on their institutional website, which solved the problem to a certain extent, if not wholly, for the students.

Furthermore, one cannot take away the fact that virtual laboratories do not prepare students for unfamiliar laboratory situations. Even though students are provided with user manuals and preparatory work on chemistry experiments, we cannot understand the "anxiety" or "fundamental scare" that student feels while entering an actual lab and performing experiments. Usually, students who may completely work on simulated chemistry experiments from the comfort of their home location may feel significant restlessness while entering the actual chemistry laboratory when the situation becomes normal. Thus, it is suggested that chemistry simulations should provide a "preview" of entering a laboratory that gives a feeling of a typical lab setting. Further, it is also suggested that teachers can create demonstration laboratories for some chemistry experiments in their campus laboratory setting wherever possible that allows students to get acquainted with the setting. A visual tour of the chemistry laboratory at the start of the semester shall give a sense of familiarity with students' future working environment, thereby building their confidence while performing experiments in an actual laboratory. In contrast to remote lab setting in robotics and electronic engineering that are fairly easy to program, chemistry lab simulation poses a challenge due to its complex requirements. The major challenge in an online chemistry laboratory is two-fold, i.e. (1) creating a virtual chemistry laboratory architecture and (2) providing a sense of "feel, see and sense" during chemical synthesis and being able to calibrate instruments for chemical analyses. For example, in UV-VIS spectrophotometric simulation, the virtual instrument should start with an initial state (zero reading). Further, simulation should provide various stages like virtually being able to hover over and drag cuvettes and test tubes and calibrate instruments. The simulation interface should allow error identification when wrong loading is done by the student. The analyte should have physical characteristics like the color (nitrobenzene and copper sulfate must look yellow and blue, respectively) on a virtual screen. During our formal interactions with students, it was revealed that providing a human "Avatar" in place of "Help Menu" in the simulation environment gives the students the feel that they are monitored, and so they are encouraged to click the avatar to ask for any assistance.

A pressing challenge was associated with cheating and dishonesty of students in virtual laboratory assessment. Most universities hold practical examinations that, if shifted to the online mode, present various issues such as use of lecture notes, utilizing more time than allocated for completing the answers, searching for answers online, or seeking help from their peers. Thus, to deter cheating and academic dishonesty, it was essential to conduct proctored examinations. Proctored examinations allow identification of students taking the test, secure login, freezing website searches while working on exam paper, and many others (Gibbs, et al. 2015). These measures are also suggested to be applied while holding practical chemistry examinations, though with flexibility of using handouts and datasheets by the students. Another challenge that was identified was teaching virtual chemistry laboratory to disabled students (Kennedy, Palmer and Boval 2013; Nepomuceno, et al. 2015). Thus, there is a need to provide virtual laboratories for disabled students in chemistry.

A rising concern was reported about whether general chemistry laboratories teach anything novel to the students or whether they were mere "cookbook chemistry" that received widespread attention in the teaching community worldwide (Rice, Thomas and O'Toole 2009). Many universities are facing financial difficulties, and it did not make sense to policymakers to provide funding for expensive chemicals, reagents, and instruments as students were not majoring in the chemistry field. This presented a serious situation for chemistry department where administrators were willing to do away with the chemistry laboratory as a whole, and the pandemic added fuel to an already existing fire. During the pedagogical design, chemistry teachers tend to misrepresent science by providing a scripted procedure to the students and orient students to favorable outcome and grades. Thus, it is suggested that rather than sending a typical "cookbook manual" to the students, conducting an argument-driven inquiry laboratory will be more useful. In an "argument-driven chemistry experiment," scheduling the same lab over multiple weeks rather than different experiment every week is suggested. Teachers can include peer assessment in lab sessions, thereby providing a dynamic learning environment for the students. Hence, it is essential to replace "old-fashioned" chemistry labs with enhanced instrumental techniques such as advanced spectroscopy, chromatography, and artificial intelligence-assisted chemistry. Chemistry experiments should be robust and relevant, and the key to this lies in their pedagogical design and successful implementation for all students.

With laboratories opening with staggered working styles, chemistry educators can conduct crucial chemistry experiments for the students in traditional chemistry labs physically while maintaining safety precautions rather than have these labs in the virtual mode entirely.

APPENDIX 1. ADVANCED INSTRUCTIONAL STRATEGIES IN VIRTUAL CHEMISTRY LABORATORY AMID THE COVID-19 OUTBREAK

Supplementary Information

Google Forms Questionnaire

Name Age Gender

1. Did you perform any experiment in physical laboratory this semester/year?

   Yes No Maybe

2. Did you feel any anxiety while performing experiments in physical laboratory in the past?

   Yes No Maybe

3. What are the experiments and issues that contributed to your anxiety while performing chemistry laboratory experimentation? ______________________

4. Did you feel simulated chemistry laboratories were convenient to use?

   Yes No Maybe

5. Did you perform simulation for aspirin preparation and analysis? Yes No Maybe

   5.1. Select all factors you experienced while performing aspirin preparation simulated experiment.

   Difficulty in Conceptual understanding Anxiety from hazards

   Procedural complexity Lack of sufficient resources

6. Have you performed polarimetric measurements of organic compounds in the simulated laboratory?

   Yes No Maybe

   6.1. Select all factors you experienced while performing polarimetric measurements of organic compound experiment.

   Difficulty in Conceptual understanding Anxiety from hazards

   Procedural complexity Lack of sufficient resources

7. Did you perform simulation experiment based on UV-VIS spectroscopy?

   Yes No Maybe

   7.1. Select all factors you experienced while performing spectroscopic simulations.

   Difficulty in Conceptual understanding Anxiety from hazards

   Procedural complexity Lack of sufficient resources

Link to Survey:
https://docs.google.com/spreadsheets/d/1Ripzl79dRFjKXKDzDRejDHg6rc8If06gf9FzvJetsKM/edit?pli=1#gid=38065717

REFERENCES

Achuthan, K., & Murali, S. S. (2015). A comparative study of educational laboratories from cost & learning effectiveness perspective. In R. Silhavy, R. Senkerik, Z. K. Oplatkova, Z. Prokopova, & P. Silhavy (Ed.), *Software Engineering in Intelligent Systems, Advances in Intelligent Systems and Computing*, vol. 349, pp. 143–153. Springer. doi:10.1007/978-3-319-18473-9_15.

Adedoyin, O. B., & Soykan, E. (2020). Covid-19 pandemic and online learning: The challenges and opportunities. *Interactive Learning Environments*. doi:10.1080/10494820.2020.1813180.

Anderson, R. M., Heesterbeek, H., Klinkenberg, D., & Hollingsworth, T. (2020). How will country-based mitigation measures influence the course of the COVID-19 epidemic? *The Lancet, 395*(10228), 931–934. doi:10.1016/S0140-6736(20)30567-5.

Bao, W. (2020). COVID-19 and online teaching in higher education: A case study of Peking University. *Human Behavior and Emerging Technologies, 2*, 113–115. doi:10.1002/hbe2.191.

Brandriet, A. R., Ward, R. M., & Bretz, S. (2013). Modeling meaningful learning in chemistry using structural equation modeling. *Chemistry Education Research and Practice, 14*(4), 421–430. doi:10.1039/C3RP00043E.

DeKorver, B. K., & Towns, M. H. (2015). General chemistry students' goals for chemistry laboratory coursework. *Journal of Chemical Education, 92*(12), 2031–2037. doi:10.1021/acs.jchemed.5b00463.

Dunnagan, C. L., Dannenberg, D. A., Cuales, M. P., & Earnest, A. D. (2020). Production and evaluation of a realistic immersive virtual reality organic chemistry laboratory experience: Infrared spectroscopy. *Journal of Chemical Education, 97*(1), 258–262. doi:10.1021/acs.jchemed.9b00705.

Flaherty, A., O'Dwyer, A., McNamara, P., & Leahy, J. J. (2017). Evaluating the impact of the "Teaching as a Chemistry Laboratory Graduate Teaching Assistant" program on cognitive and psychomotor verbal interactions in the laboratory. *Journal of Chemical Education, 94*(12), 1831–1843. doi:10.1021/acs.jchemed.7b00370.

Galloway, K. R., & Bretz, S. L. (2015). Development of an assessment tool to measure students' meaningful learning in the undergraduate chemistry laboratory. *Journal of Chemical Education, 92*(7), 1149–1158. doi:10.1021/ed500881y.

Gamage, K. A., de Silva, E. K., & Gunawardhana, N. (2020). Online Delivery and Assessment during COVID-19. *Education Sciences, 10*(11), 301. doi:10.3390/educsci10110301.

Gibbs, H. C., Gupta, N., Northcutt, C., Cutrell, E., & Thies, W. (2015). Deterring cheating in online environments. *ACM Transactions on Computer-Human Interaction, 22*(6), 1–23. doi:10.1145/2810239.

Harle, M., & Towns, M. (2011). A review of spatial ability literature, its connection to chemistry, and implications for instruction. *Journal of Chemical Education, 88*(3), 351–360. doi:10.1021/ed900003n.

Hofstein, A., Nahum, T. L., & Shore, R. (2001). Assessment of the learning environment of inquiry-type laboratories in high school chemistry. *Learning Environments Research, 4*(2), 193–207. doi:10.1023/A:1012467417645.

Jones, N. (2018). Simulated labs are booming. *Nature*, 562(7725), S5–S7. doi:10.1038/d41586-018-06831-1.

Jong, T., Linn, M. C., & Zacharia, Z. C. (2013). Physical and virtual laboratories in science and engineering education. *Science, 340*(6130), 305–308. doi:10.1126/science.1230579.

Josephsen, J., & Kristensen, A. K. (2006). Simulation of laboratory assignments to support students' learning of introductory inorganic chemistry. *Chemistry Education Research and Practice, 7*(4), 266–279. doi:10.1039/B6RP90013E.

KANTAR IMRB. (2019). *ICUBE™: Internet In India*. Survey, Mumbai.

Kennedy, S., Palmer, J., & Boval, J. (2013). Adaptive tools for laboratory students with disabilities: Case studies from UC San Diego. *Journal of Chemical Health and Safety, 20*(1), 35–42. doi:10.1021/acs.chas.8b20109.

Kolb, D. A., Boyatzis, R. E., & Mainemelis, C. (2001). Experiential learning theory: Previous research and new directions. In R. J. Sternberg, & L.-F. Zhang (Eds.), *Perspectives on Thinking, Learning, and Cognitive Styles* (p. 227). Routledge. doi:10.4324/9781410605986-9.

Murphy, M. (2020). COVID-19 and emergency e Learning: Consequences of the securitization of higher education for post-pandemic pedagogy. *Contemporary Security Policy, 41*(3), 492–505. doi:10.1080/13523260.2020.1761749.

Nakhleh, M. B., Polles, J., & Malina, E. (2002). Learning chemistry in a laboratory environment. In J. K. Gilbert, O. D. Jong, R. Justi, D. F. Treagust, & J. Van Driel (Eds.), *Chemical Education: Towards Research-based Practice* (pp. 69–94). Kluwer Academic Publishers. doi:10.1007/0–306–47977-X_4.

Nam, T. (2019). Understanding the gap between perceived threats to and preparedness for cybersecurity. *Technology in Society, 58*, 101122. doi:10.1016/j.techsoc.2019.03.005.

Nepomuceno, G. M., Decker, D. M., Shaw, J. D., Boyes, L., Tantillo, D. J., & Wedler, H. (2015). The value of safety and practicality: Recommendations for training disabled students in the sciences with a focus on blind and visually impaired students in chemistry laboratories. *Journal of Chemical Health and Safety, 23*(1), 5–11. doi:10.1016/j.jchas.2015.02.003.

Piaget, J. (1964). Cognitive development in children: Piaget development and learning. *Journal of Research in Science Teaching, 2*, 176–186. doi:10.1002/tea.3660020306.

Plunkett, K. N. (2019). A simple and practical method for incorporating augmented reality into the classroom and laboratory. *Journal of Chemical Education, 96*(11), 2628–2631. doi:10.1021/acs.jchemed.9b00607.

Reid, N., & Shah, I. (2007). The role of laboratory work in university chemistry. *Chemistry Education Research and Practice, 8*(2), 172–185. doi:10.1039/B5RP90026C.

Rice, J. W., Thomas, S. M., & O'Toole, P. (2009). *Tertiary science education in the 21st century.* The Australian Learning and Teaching Council.

Rosário, P., Núñez, J. C., Vallejo, G., Cunha, J., Nunes, T., Suárez, N., Fuentes, S., Moreira, T. (2015). The effects of teachers' homework follow-up practices on students' EFL performance: A randomized-group design. *Frontiers in Psychology, 6*(1528), 1–11. doi:10.3389/fpsyg.2015.01528.

Roy, B., & Roy, A. (2020). Conducting examinations in India: Emergency, contention and challenges of students amidst covid-19 pandemic. *Children and Youth Services Review, 120*, 105768. doi:10.1016/j.childyouth.2020.105768.

Sutton, H. (2020). Keep your mission student-centered, even in the face of crisis. *Dean & Provost, 21*(10), 12. doi:10.1002/dap.30738.

Tatli, Z., & Ayas, A. (2010). Virtual laboratory applications in chemistry education. *Procedia Social and Behavioral Sciences, 9*, pp. 938–942. doi:10.1016/j.sbspro.2010.12.263.

UNESCO. (2020). *COVID-19 educational disruption and response.* Retrieved December 30, 2020, from https://en.unesco.org/covid19/educationresponse.

5 Learner's Autonomy in a Weblog-Mediated ESL Class in Pakistan

Sahib Khatoon Thaheem and Habibullah Pathan
Mehran University of Engineering and Technology

Mohammad Jafre Zainol Abidin
University Sains Malaysia

Quratulain Mirza
Mehran University of Engineering and Technology

CONTENTS

5.1 INTRODUCTION

Computer-Assisted Language Learning (CALL) has grown rapidly within a short period. Teachers who teach face to face are called on to adapt their courses to the online environment (Kreijns et al. 2014). In this era, the way of teaching-learning is changed all over the globe; thus, the need in Pakistan right now is to also change its learning and teaching environment from the conventional class to a self-directed

DOI: 10.1201/9781003143796-5

class that is easier for the learners and makes them free to learn and enjoy simultaneously because learners always need their own space (Van Dozer 1997). The evolving technology has made this world a global village, and this connectivity is not space-bound, thus making learning accessible and easy (Khatoon Thaheem et al. 2016). According to Majeski, Stover, and Valais (2018), adult learners are filtered learners who filter themselves from surroundings and are much motivated, so they can take advantage of CALL easily. The digital revolution must be accepted and the teaching-learning strategies with the use of different technological tools must be developed and updated (Mahboob 2017). Tafazoli, Parra, and Abril (2017) opined that the innovative digital tools existing today play a significant part in our routine. Nowadays, if anyone is using the latest gadgets and tools in their common chores, either at the home, office, or in education, then they are considered an updated and digital being of today's world. There are numerous technological tools (e.g., wiki, blog, podcast, social bookmarking, social networking sites, broadcasting sharing sites, cloud computing, etc.) that have enabled out-of-the-box transformation of the teaching purposes (Toffoli and Sockett 2015). This persistent existence of technology integration creates unlimited opportunities for its use in various fields.

Weblogs offer the learners wide-ranging chances of learning in communicative, collaborative, and immersive ways (Genzola 2015). According to Harvil (2018), blogging is an interactive platform of language learning that can be used to improve communication skills. Wang and Hsua (2008) stated that blogs are a very flexible tool that offers a richer learning environment than a traditional classroom. It has multiple functions that can improve the class environment as blog entries can be cited or linked by the public, categorized, and archived after the class, helping instructors connect one class to another. Weblog offers learners with various choices and changes their learning strategies, thus supporting autonomy expansively (Bhattacharya and Chauhan 2010) because learning is not possible in a vacuum and "independence" does not automatically infer learning on one's own; communication, cooperation, collaboration, and connectivity promote learning and learner's autonomy (Bhattacharya and Chauhan 2010). Teachers can increase students' confidence and autonomy with planned teaching strategies and by integrating technology that students follow confidently (García and Arias 2010); also, student must perform independently and be able to troubleshoot. Thus, the numerous platforms are giving confidence to students, and those platforms create a student-centered environment (Khatoon 2018). Thus, technology integration in ESL teaching and learning makes students self-directed, confident, and autonomous (Thaheem et al. 2021; Khatoon et al. 2020). The current study explores whether the autonomy of learners can be enhanced in the digital context of (Weblog) mediated class or not.

These subsequent objectives are attained:

1. To explore if the weblog-mediated teaching can enhance learners' autonomy.
2. To identify the significant difference between girls' and boys' autonomy in the (CALL)Weblog class.
3. To explore students' perceptions about the weblog-mediated class by their postings on the weblog during study.

5.2 RELATED LITERATURE

Computer-assisted learning (CALL) is not new, and has been for over 40 years in the education system globally, and there is an overabundance of publications dedicated to exploring its effect and implications for practice. Thus, the field is deep-rooted, with a bulk and ever-growing practice of research and inferences (Jarvis 2013). Ashraf (2016) posited that it is required for English language teachers in Pakistan to become competent in using computers in EFL teaching and learning in this digital era. Khatoon Thaheem et al. (2016) mentioned that CALL and technology can benefit to boost the learner's confidence.

5.3 SNAPSHOT HISTORY OF ONLINE EDUCATION

The effect of technology was determined in the ventures between Apple and different universities that were called the Apple classrooms of Tomorrow (ACOT) (Beisel et al. 2017) in a project initiated in 1985; this was what led to the transition from traditional classrooms to classrooms having computer usage as an instructional tool in 1995. That project was having to explore a few things: (1) Students' activeness for the exploration of new ideas, rather than passive acceptance; (2) Work based on group activity and peer-based tasks, rather than individually; and (3) Usage of technology to solve real-world problems. During this project, teachers started shifting their traditional classroom into technology-based classrooms and identified that this integration was beneficial for students' learning and engagement (Beisel et al. 2017). There are tools, such as flashcards, charts, models, tape recorders, computers, scripts, films, overhead projectors, pictures, and audio-visual aids, which are present to conduct online education, but they are not used as they should be (Abdul, Awan, and Shafi 2016). According to Malcolm et al. (2008), online learning has the advantage to save time and availability; students can customize their program and structure according to their ease. Either synchronous or asynchronous they can discuss also, thus this learning is not time, or venue bound. With the expansion of IT and the internet, which has almost changed human life in many ways, including a paradigm shift in education, many universities are using e-learning additionally to conventional learning globally (Berliyanto and Santoso 2016).

In Pakistan, there is a dire need for this shift, and the higher education commission (HEC) began an English Language Teaching Reforms in 2004 to upgrade English faculty by continuous professional development and CALL.

5.4 LEARNER'S AUTONOMY

Autonomy refers to when people take charge to control over their doings and when students' charge and control over their learning in classrooms or out of class with purpose. The term self-directed learning,' also called 'independent learning,' also directs you to learn by yourself (Wenden 1997). In the wake of the 1990s, autonomy became the central focus in the classroom. With this significance in the classroom, the role shifting also took place from a teacher-centered to a learner-centered class, which meant the control was shifted from teachers to students (Richards 2015). Furthermore, the

Vygotszkian collaborative and social-learning approach and the change of student–student interdependence also be stretched out of the classroom by making the learner fully independent (Nunan and Richards 2015; Murray 2014; Reinders 2018). For being autonomous in the class or out of the class, learners needs to be active, able to work their own, and, most important, must have a willingness to learn independently or with teachers and peers (Wilson 2018). A learner who is a trouble-shooter, initiator, independent, and willing to learn will be autonomous, but there should be an interesting context for him to activate his motivation and willingness and it comes with teachers' teaching methodology as well; thus, if the material, context, and methodology are not interesting for the student, they will follow the tasks unwillingly or may not follow these at all, resulting in a loss of willingness (Adams, Little, and Ryan 2017).

5.5 LEARNERS' AUTONOMY IN TECHNOLOGY-INTEGRATED ESL/EFL CLASS

Since the launch of technology 20 years ago, the relationship between language learning and technology and autonomy has progressed and become promising (Reinders and White 2016). Starting with Little (2001) was the notion of telecollaboration at the beginning of the broadband era, followed the developments in (ICT) over the last 10 years, and mainly Web 2.0 revolution (Thorne 2014), which led to the continuous interconnectivity of people for the exchange of information, ideas, and practices; it extended the classroom-based social-learning to a greater extent (Reinders and White 2016). According to Reinders and White (2016), learner autonomy has become important and has become the goal of language learning context globally. Encouraging learner autonomy remains the major concern for instructors and researchers. Ardi (2017) states that learner autonomy is the learners' "ability to be responsible for his doings by setting goals and owing to the learning process by monitoring that". The model in Figure 5.1, by Reinders and White (2016), represents this development across the last two decades with increasing rates of change being evident.

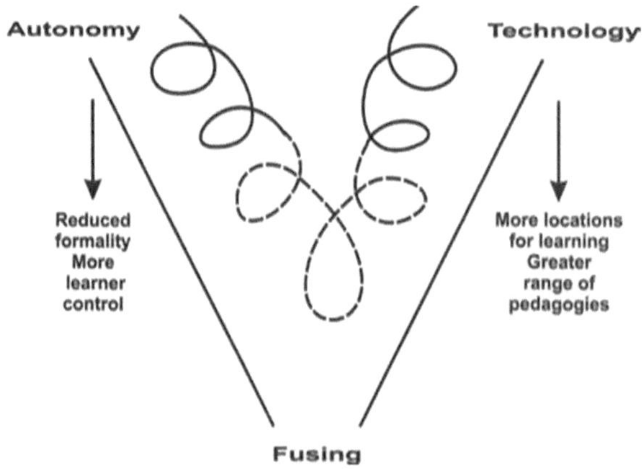

FIGURE 5.1 A model of convergence.

Web 2.0 technologies are supposed to offer an autonomous educational platform that reduces the authority structure between the instructor and learners, encouraging the learners to show their capacity to direct and control their learning (Ardi 2017). Similarly, technology encourages autonomy as learners have the liberty to manage their learning and cognitive processes. As, Alm (2006) posited, CALL context can enhance learners' interest in learning the tasks that show their improvement in autonomy and provide them with a feeling of identity. According to Tassinari and Berlin (2012), autonomy comprises the learner's active involvement in language learning along with the control over the learning material. CALL materials are offered to the learners without any limitation of time and space, which motivates students to learn the language out of the class easily and independently without any teachers' constant instructions or guidance. CALL materials provide quick feedback, which also makes the student autonomous and self-reliant; it reduces the dependence on the teacher. To foster learner autonomy, learning resource provision must be according to the student's level and understanding of the work with them in or out of class (Richards 2015; Wenden 1997). Since the computer facilitates monitoring learners' behavior and improvement, it can modify the input, or make recommendations to the learners, based on their performance; these records help students to monitor their progress, make decisions about the learning, and make them responsible to reflect and form priorities (Reinders 2010). According to Lin and Reigeluth (2019), for scaffolding learner autonomy, learners' mindsets have to change; for this, careful instructional designs are required to support the behavioral, cognitive, and affective aspects to decrease the resistance.

The plethora of research has also documented the strong relationship between technology and learner autonomy (Hazaea and Alzubi 2018; Reinders 2010; Lee 2011). The research conducted by Pasaribu (2020) indicated that digital reader response activities in a Moodle-based reading class empowered learners to plan, implement, and assess their learning. The findings revealed online context and tasks not only encouraged their motivation toward the tasks but also nurtured and enhanced their social dimensions of autonomy. Other research conducted by Hazaea and Alzubi (2018) reported that in the reading context, they used WhatsApp, which developed the sense of learner autonomy because of its reflective and connectivity characteristics; thus, students learnt reading even out of the class on WhatsApp. The action research was undertaken at a university language center in Ambato, Ecuador, by Edwards, Holguín-barrera, Ortiz and Pérez (2019), with the learners of English as a Second Language (EFL), and the findings exhibited the development of learner autonomy with the platform-based collaborative learning approach. Oussou (2020) investigated the degree of autonomy that Moulay Ismail University EFL students enjoyed and the extent to which they used ICT tools for their learning and investigated the relationship between autonomy and ICT tools how ICT, the findings of the study revealed positive relation between both, autonomy and ICT use. ICT use is validated by the current study.

5.6 WEBLOG AND ITS IMPLICATIONS IN ESL

As a brief history, blogging commenced in about 1993 as a medium for the National Centre for Super Computing Applications (NCSA). The term "weblog" was voiced for the first time in 1997 by an internet fan called Jorn Barger. He decided to rename his website (Robot Wisdom) a "weblog," which denotes websites that are being updated nonstop (Ward 2004). Weblogs are declared to have probable standards in English language learning for several reasons and this context provides chances for real communication in an authentic learning context (Noytima 2010). Weblogs offer communicating and expressive interface in the target language because learners can publish and can express, post, and comment, which enables them to select their own time for engagement in any topic of their interest as an after-class language exercise platform for the enhancement of their target language skills (Genzola 2006). Three forms of Weblog can be used for language learning, i.e., the learner Blog, the tutor Blog, and the class Blog. On learner blogs, learners post their ideas, views, and thoughts related to the topic; therefore, it promotes self-expression of the learners and promotes the sense of owing their expressions, leading learners to attain autonomy. The tutor blog is used by the teachers for students, and teachers keep their material related to English websites and resources, which students can link with, for self-access. The class blog boosts learners' autonomy to post messages, images, and links vis-a-vis classroom dialogues and the topics (Noytima 2010; Lee 2011; Khatoon et al. 2020). Individual blogs used by students or teachers foster self-expression and self-reflection, although collective blogs connecting a whole class or small groups encourage collaborative and combined learning (Lee 2011). The Weblog-mediated class context encourages individuals to discuss the learning beyond the class; there every learner is free and can take charge to initiate their discussion by posting their thoughts and can and give and receive feedback (Alm 2006). Thus, Mynard (2007) suggests that the Weblog tool should be used by the instructors to encourage students' reflective sense, and learners would be autonomous by deciding when, how, and what to publish and comment on (Lee 2010). The researchers globally have investigated how weblog is beneficial or not in language learning; Nezakatgoo and Tabatab (2019) investigated, by explanatory sequential design, with 46 tertiary students, the learners' autonomy in weblog-mediated writing instruction class. The findings of the study revealed that learner's autonomy (cognitive and metacognitive) was enhanced under weblog-mediated writing class along with their writing skill. Besides, the course was incorporated on a weblog with a dictionary, and other vocabulary concerned websites were linked (Shams 2013) for testing vocabulary and learner autonomy in a study. The findings of the research showed very positive effects on learners' autonomy and vocabulary knowledge. Another research conducted at the University of Thailand by Noytima (2010) in the EFL context using weblog showed that students' motivation, leadership skills, writing skills, critical thinking, and sense of readership and authorship increased significantly. Besides, in Malaysia, the study investigated the ESL students' perceptions and their experiences in the weblog-mediated writing skill class, in which writing autonomy was a focused objective; the findings of the study revealed students not only enjoyed the working weblog but they felt weblog-mediated classed gave them the opportunity of authorship and publishing feature related to their interest freely, thus promoting their autonomy (Foroutan 2013). The study research conducted by Genzola (2006) on the use of weblog in an educational institute of China was investigating learner autonomy and university learners' collaborative learning with a sample of 108 students who reflected about their weblogging experience, which was quite positive; the respondents stated that their autonomy and collaborative learning enhanced in the weblog-mediated class. According to Genzola (2006), language teachers/instructors must focus on learning context and on the curriculum to ensure it is tailor-made them according to the needs of the students to make them innovative, conducive, competitive, and self-motivated, critical learners. One more study conducted by Khatoon et al. (2020), on blogging in the ESL context with respect to gender differences in their attitudes toward technology-mediated (blogging) platform found very few differences among male and female students toward using blog in their ESL class; overall, the impact was positive, and students positively adopted this medium to improve their language skills. Hence, the current study is investigating the students' perceptions regarding the newly designed program of ESL on the Weblog to assess if they feel comfortable with this innovation or not.

5.7 WEBLOG CHAT POSTINGS

Teachers in the educational sector found it worthwhile in the United States in the teaching of writing skills, where they discovered that computer-mediated communication had nearly the same impacts as face-to-face classes (Warschauer 1996). Chatrooms or placing comments on any platform give a way to learners to have real communication online out of the class, in either oral or text form, according to their ease and convenience (Richards 2015). According to Macintyre (2007), chatroom is providing a restriction-less environment for English language learners to comment and communicate in a stress-free manner so that they do not bother about any mistake, even if they make one; thus, their interaction becomes their base for learning communication and comprehension more than classroom-based communication (Jenks and Jenks 2010). Besides, Jepson (2005) opined that chat room contact might not only increase understanding of the language learners use, providing them with the opportunities for self-correction and making their identity as an ESL learner, in in-text chat rooms, language learners communicate in real time on their laptops/ phones/ computers by using the internet by exchanging messages. Those chats are real time and also have errors similar to that seen in face-to-face interaction and is just as informal and carries the same advantages of language learning as face-to face-language learning (Smith 2003; Warschauer 1996).

5.8 METHOD

Keeping in mind the structures of the current study, the survey method was followed to collect the data from five departments of Mehran UET Jamshoro through the questionnaire and postings/ comments (chat box of weblog) of the students. The research approach is a mixed methods approach in which the QUAN=QUAL approach is followed. The sample of the study included students from the Electrical, Electronics, Computer System, Software, and Mechanical engineering department who were taught using Weblog (the designed program) in the first semester of the first year.

5.9 DATA MANAGEMENT

The designed questionnaire was consisted of 20 items to know the autonomous behavior of the learners in the (CALL) Weblog class.

5.10 PILOT STUDY

A pilot study was conducted with 25 respondents (5 from each department) learning through a weblog-mediated class in the Mehran UET Jamshoro. An Item Reliability Analysis was run to check the reliability of the questionnaire, and the value for the "Cronbach's alpha reliability test" was 0.728.

5.11 RELIABILITY OF THE QUESTIONNAIRES

The Alpha value was between 0 and 1. The Alpha value should be at least 0.70. This is a significant value and therefore, the tool is reliable. Table 5.1 shows the reliability of the current study tool (questionnaire).

5.12 DATA COLLECTION PROCEDURE

For the quantitative analysis, using the questionnaire, data were collected from 150 first-year students of five departments who were taught English subjects through the Weblog-mediated platform. Among the 150 students, there were 88 male participants and 62 female participants. After quantitative data collection, students' postings/comments were taken from the weblog where students interacted with each other and gave comments. The comments of the participants were coded; based on those codes, thematic codes were drawn. The data of all respondents were dependent on examination and organization procedures to discover commonalities, and codes were derived from that. This application is for simple arrangement and categorization of comments; thus, for the reinforcement of the qualitative analysis. According to Ivankova, Creswell, and Stick (2006), "qualitative data analysis mainly requires categorizing things, people, and proceedings and the properties which characterize them". After data collection, it was coded, systematized, planned, and investigated to identify the answer to the research query.

5.13 DATA ANALYSIS AND RESULTS

In response to the first and second objective of the research, inferential statistics were completed in which the descriptive analysis was done to achieve the first objective

TABLE 5.1
Reliability of the Questionnaire

Cronbach's Alpha	Cronbach's Alpha Based on Standardized Items	N of Items
0.723	0.727	20

and then the Independent-Samples t-test was run in SPSS to determine whether there were any statistically significant differences in the perceptions of the male and female students vis-à-vis autonomy enhancement in the CALL (Weblog mediated) context. As Pallant (2010) states, independent-samples t-test measures the difference between two groups on numerous hypotheses. To answer the third objective, content analysis was done by the researcher by analyzing the students' postings on the Weblog.

Table 5.2 shows the participants and the percentages given by the participants to each item, examining the learner's autonomy.

The sample had more male respondents ($n=88$) than females ($n=62$).

Male and female's perceptions were almost similar with regard to the autonomy. Nevertheless, some differences were found where males and females disagreed each other. Within the female and male groups, the standard deviations are relatively close. The questionnaire items were individually analyzed by the researcher using descriptive analysis to know areas of agreement and disagreement between the students because there are four items in the questionnaire that negate the Weblog-mediated class.

The results of the first objective of the study revealed that students strongly admitted that they want to learn in CALL context, and they feel relatively autonomous and feel joy in learning via a CALL (Weblog)-mediated class. To achieve the second objective of the study, the quantitative analysis was done to examine the significant difference between two groups (male and female) to see the differences in perceptions if any. Table 5.3 shows these results.

An independent-samples *t*-test was conducted to investigate if there were any significant differences between the genders in terms of responses regarding their autonomy in the weblog-mediated ESL class. Table 5.3 shows that there was no significant difference between boys (M=59.1818, SD=8.67358) and girls (M=59.5645,

TABLE 5.2
Descriptive Analysis

			Autonomy		
Gender	Mean	N	Std. Deviation	Minimum	Maximum
Male	59.1818	88	8.67358	24.00	72.00
Female	59.5645	62	6.27412	49.00	75.00
Total	59.3400	150	7.75102	24.00	75.00

TABLE 5.3
Independent-Samples *t*-Test

	BOYS		GIRLS				
Learner Autonomy	Mean	*SD*	Mean	*SD*	*t*	*df*	*p*
	59.1818	8.67358	59.5645	6.27412	314	148	.767

SD = 6.27412) in terms of their autonomy in the weblog-mediated ESL class, t (314) =, $p = .767$

The above-mentioned results showed a very positive effect of the weblog-mediated context on students' autonomy as they agreed with ideas given in the questionnaire that they enjoyed, felt free, and resourceful, comfortable, self-directed, and self-monitored themselves while working and learning in the CALL context (weblog-mediated ESL class). On weblog, students were giving feedback by clicking the stars; the majority clicked on 5 stars, which shows how comfortable they were in the given context (Figure 5.2).

These results suggest that boys and girls had enhanced autonomy equally in the weblog-mediated ESL class; thus, no gender-related differences exist in the Weblog-mediated context and autonomy enhancement. To achieve the third objective of the study, researchers analyzed and recorded the postings of the students when they commented about their experience about the weblog-mediated class. Those comments were qualitatively analyzed with content analysis by following the steps mentioned below (Table 5.4).

Decontextualization
Recontextualization
Categorization
Compilation

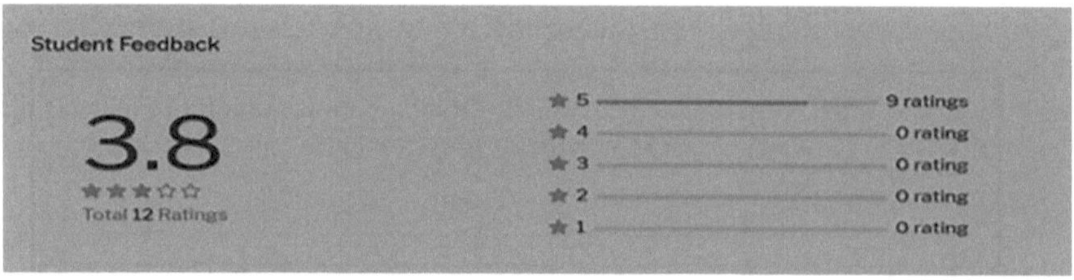

FIGURE 5.2 Students' feedback.

TABLE 5.4
Derived Themes

Autonomy	Themes
Monitoring	Energetic
Motivated	Responsible
No limitations	Self-enough
Liberty	Self-directed
Supportive	Without support
Creative	Helpful
Interesting	Authority
Authorization	No restrictions
Self-driven	Free
Freedom	Student-centered
Choice	Preference

Students' postings' regarding quick online feedback displayed on the weblog are mentioned below in Figures 5.3 and 5.4.

According to Tassinari and Berlin (2012), online quick feedback enhances the motivation of the students to learn and work (Figure 5.5).

In line with the opinion of Malcolm et al. (2008), online learning is not time, or venue bound; so the student works according to his ease because weblog is present all the time; whenever they want to work, they will sign in and complete their task either on their laptops or mobile phones. Thus, students feel free to work and learn in their own time (Figures 5.6 and 5.7).

In line with Genzola's (2006) findings, students enhanced their autonomy and learned the language as well in a collaborative, online connected environment. It was also shown that students are enjoying and learning in online communication (Figures 5.8 and 5.9).

According to Shams (2013), the addition of a dictionary in the weblog for teaching vocabulary enhanced students' autonomy; similarly, in this study too, in the comments, students expressed that their motivation increased because of videos and other

FIGURE 5.3 Students' feedback.

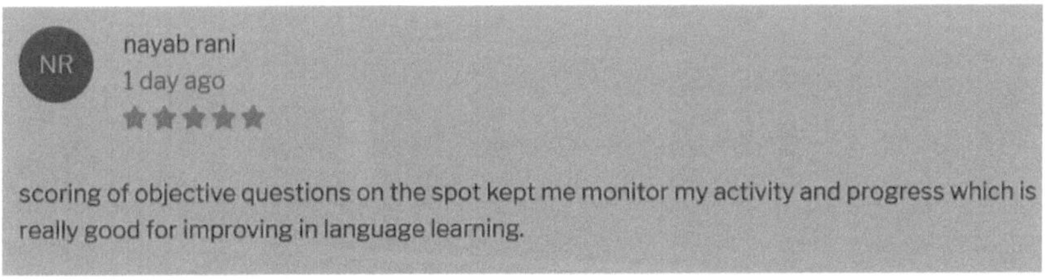

FIGURE 5.4 Students' feedback.

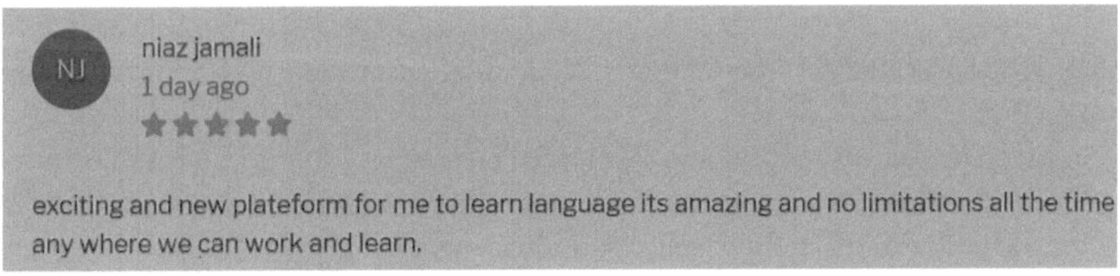

FIGURE 5.5 Students' feedback.

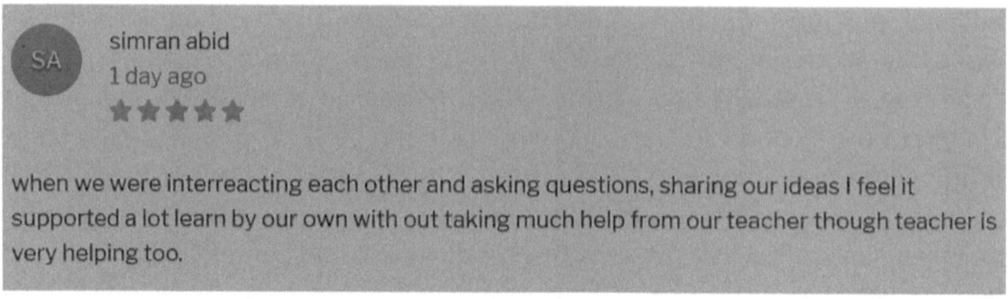

FIGURE 5.6 Students' feedback.

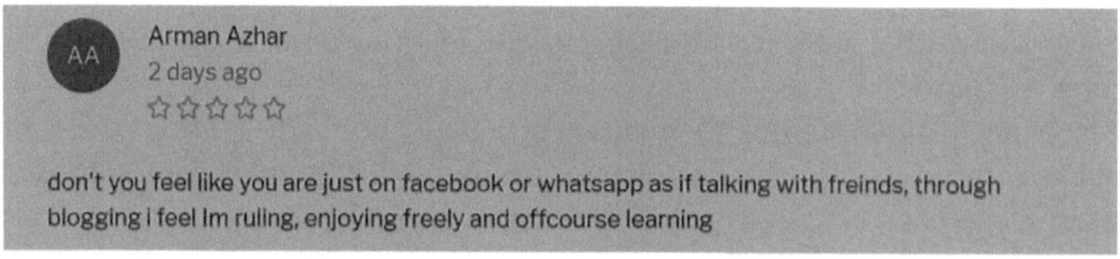

FIGURE 5.7 Students' feedback.

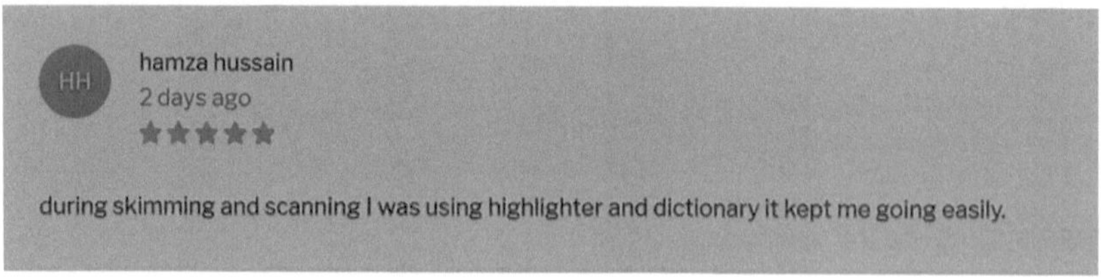

FIGURE 5.8 Students' feedback.

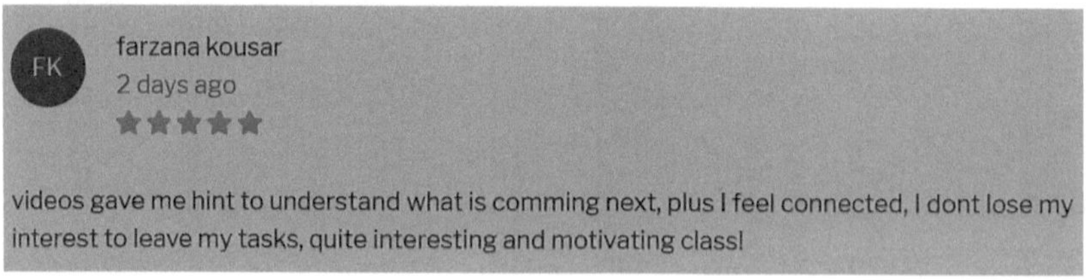

FIGURE 5.9 Students' feedback.

features of the weblog. So, the students are feeling motivated to work and continue learning (Figure 5.10).

According to Adams, Little, and Ryan (2017), learners should be troubleshooters and willing to initiate the task, the above-mentioned comments by the respondents show they feel have sufficient resources and can work their own, which is a sign of autonomy (Figure 5.11).

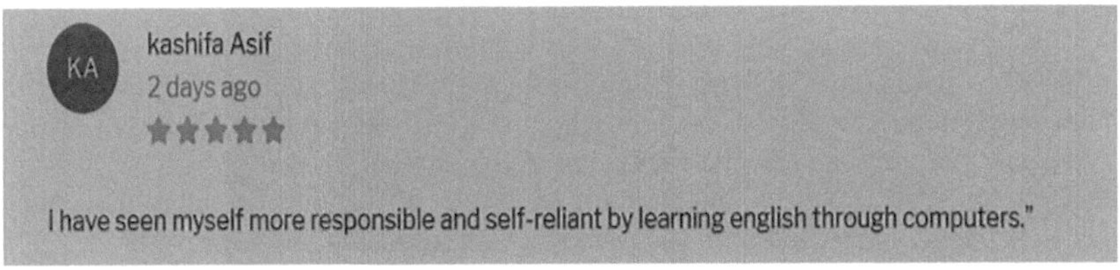

FIGURE 5.10 Students' feedback.

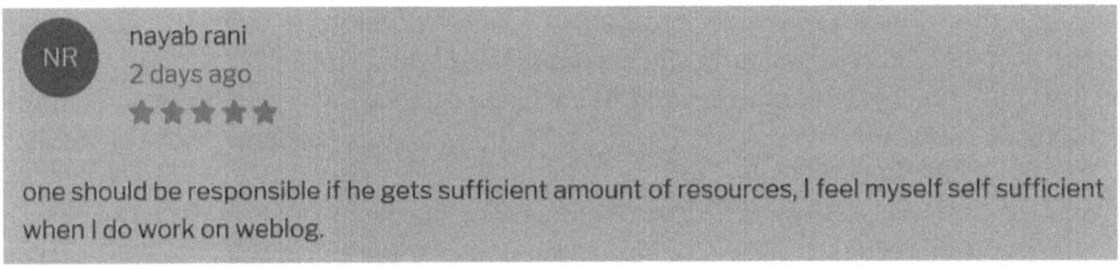

FIGURE 5.11 Students' feedback.

5.14 CONCLUSION

This study investigated the students' perceptions regarding weblog-mediated class: either they felt autonomous or not in the CALL context. Overall the findings showed their positive attitude toward weblog-mediated learning because they feel free, self-directed, motivated, and self-sufficient even without a teacher, which reveals they feel autonomous. Blogging encouraged their self-directed learning; as they were posting and they were connected with their peers, they felt they were connected and they are learning by that communication as well. Data analysis indicated that there was no significant difference among the perceptions of boys and girls in terms of autonomy. As the mixed-method approach was applied to collect and assess the data which can vividly reflect students' opinions, the results showed that students were improving their language learning and autonomy as well.

The current research can be considered a groundbreaking research regarding weblog-mediated ESL class because, as yet in Pakistan, weblog-mediated courses or classes are not organized for the students, despite the fact that CALL is not employed in educational settings such as schools, colleges, and universities in Pakistan because it is still in the embryonic stage but it should be used as it is used around the globe. As in this digital age students must learn under the environment that makes them self reliant and self-sufficient to work their own, CALL provides such a facility; thus, instructors and policymakers and the designers of the curriculum must focus to bring technology into their plans and strategies such that teachers can bring these in their methodology. Also, this study can open the door for course designers, researchers of Pakistan, to bring this idea into consideration and work on this to bring change in tertiary education system, especially in ESL classes.

REFERENCES

Abdul, Ghafoor Awan, and Mehvish Shafi. 2016. “Analysis of teaching methods of english language at government secondary school level in dg khan city-Pakistan.” *Global Journal of Management and Social Sciences* 16(8): 9–17.

Adams, Nicole, Todd D. Little, and Richard M. Ryan. 2017. “Self-determination theory.” *Development of Self-Determination Through the Life-Course* 55(1):47–54.

Alm, Antonie. 2006. “CALL for autonomy, competency and relatedness: Motivating language learning environments in web 2.0.” *The JALT CALL Journal* 2(3):29–38.

Ardi, Priyatno. 2017. “Promoting learner autonomy through schoology M-learning platform.” *Teaching English with Technology* 17(2):55–76.

Ashraf, M. 2016. “The effectiveness of integrating computer assisted language learning (CALL) into the textbookbased syllabus of english language: teachers’ and learners’ perspectives.” *Research Journal of Language & Linterature* 1(1):115–140.

Beisel, Carolyn, Review Committee, Committee Chairperson, Education Faculty, Committee Member, Education Faculty, Education Faculty, Chief Academic Officer, and Eric Riedel. 2017. “Walden University.”

Berliyanto, and Harry B. Santoso. 2016. “Indonesian perspective on massive open online courses: Opportunities And.” *Journal of Educators Online*. doi: 10.9743/JEO2018.15.1.11.

Bhattacharya, Atanu, and Kiran Chauhan. 2010. “Augmenting learner autonomy through blogging.” *ELT Journal* 64:1–9.

Edwards, Roger, Holguín-Barrera, María Daniela, Ortiz, Ana Cristina, & Pérez, Mariela. (2019). “Promoting EFL learner autonomy in a teacher-centered culture through video-sharing and collaborating in online forums.” *Latin American Journal of Content & Language Integrated Learning* 12(1): 99–127.

Foroutan, Maryam. 2013. “Weblog promotes ESL learners’ writing autonomy.” *Journal of Language Teaching and Research* 4(5):994–1002.

García, Mercedes Rico, and Filomena Vinagre Arias. 2010. “A comparative study in motivation and learning through print-oriented and computer-oriented tests.” *Computer Assisted Language Learning* 13:8221.

Genzola, Arnel E. 2006. “Learner autonomy through weblogging: An ICT-mediated pedagogy in ELT.” *Frontiers of Language and Teaching*, 1:4–20.

Genzola, Arnel E. 2015. “Weblogs on language learning: A technology-enhanced instruction in a tertiary-level EFL classroom in china arnel.” *Arab World English Journal (AWEJ)* 6(4). doi:10.2139/ssrn.2847467.

Harvil, Holly Renee. 2018. “Teacher Perceptions on the Use of Technology With English.”

Hazaea, Abduljalil Nasr, and Ali Abbas Alzubi. 2018. “Impact of mobile assisted language learning on learner autonomy in EFL reading context.” *Journal of Language and Education* 4(2):48–58.

Ivankova, Nataliya V., John W. Creswell, and Sheldon L. Stick. 2006. “Using mixed-methods sequential explanatory design: From theory to practice.” *Field Methods* 18(1):3–20.

Jarvis, Huw. 2013. “Computer assisted language learning (CALL): Asian learners and users going beyond traditional frameworks.” *Asian EFL Journal* 15(1):190–201.

Jenks, Chris, and Chris Jenks. 2010. “Adaptation in online voice-based chat rooms: Implications for applied linguistics.” 147–148.

Jepson, Kevin. 2005. “Conversations – and negotiated interaction – in text and voice chat rooms.” *Language, Learning and Technology* 9(3):79–98.

Khatoon, Sahib. 2018. “Level of autonomy in computer assisted language learning (Call) class; A study of engineering students (Call) Class.” (December 2017).

Khatoon, Sahib, Muhammad Jafre, Zainol Abidin, Quratulain Mirza, and Ashfaq Hussain. 2020. “Blogging in ESL class- gender-based attitude of the engineering students.” *International Journal of Evaluation and Research in Education (IJERE)* 9(4):1128–37.

Khatoon Thaheem, Sahib, Quratulain Mirza, and Shoukat Ali Lohar, 2016. "Investigating internet influence in english class." *Eurasian Journal of Humanities Computer Assisted Language Learning (CALL)* 1(2): 54–63.

Kreijns, Karel, Frederik Van Acker, Marjan Vermeulen, and Hans Van Buuren. 2014. "Community of inquiry: Social presence revisited." *E-Learning and Digital Media* 11(1):5–18.

Lee, Lina. 2011. "Blogging: Promoting learner autonomy and intercultural competence through study abroad." *Language Learning and Technology* 15(3):87–109.

Little, B. R. (2001). Personality psychology: Havings, doings, and beings in context. In J. S. Halonen & S. F. Davis (Eds.), The many faces of psychological research in the 21st century Society for the Teaching of Psychology.

Lin, Chun-Yi, and Reigeluth, Charles M.. (2019). "Scaffolding learner autonomy in a wiki-supported knowledge building community and its implications for mindset change." *British Journal of Educational Technology* 50(5), 2667–2684.

Macintyre, Peter D. 2007. "Willingness to Communicate in the Second Language : Understanding the Decision to Speak as a Volitional Process."

Mahboob, Ahmar. 2017. "English Medium Instruction in Higher Education in Pakistan : P Olicies, Perceptions, Problems, and Possibilities." In: Fenton-Smith B., Humphreys P., Walkinshaw I. (eds.) *English Medium Instruction in Higher Education in Asia-Pacific. Multilingual Education*, vol. 21. Springer, Cham, 71–92. doi:10.1007/978-3-319-51976-0_5.

Majeski, Robin A., Merrily Stover, and Teresa Valais. 2018. "The community of inquiry and emotional presence." *Adult Learning* 29(2). doi:10.1177/1045159518758696.

Malcolm, Campbell, Will Gibson, Andy Hall, David Richards, Peter Callery. 2008. "Online vs face-to-face discussion in a web-based research methods course for postgraduate nursing students." *Elsevier Editorial System(Tm) for International Journal of Nursing Studies* 45: 750.

Murray, IGarold. 2014. *Social Dimensions of Autonomy in Language Learning.*

Mynard, Jo. 2007. "A Blog as a Tool for Reflection for English Language Learners." (November):1–10.

Nezakatgoo, Behzad, and Allameh Tabatab. 2019. "Second language writing through blogs: An investigation of learner autonomy." *Iranian Journal of Applied Language Studies* 11(2):165–90.

Noytima, Usa. 2010. "Weblogs enhancing EFL Students' english language learning." *Procedia - Social and Behavioral Sciences* 2:1127–32.

Oussou, Said. (2020). "Promoting Learner Autonomy: The case of Moroccan secondary EFL teachers." *International Journal of Language and Literary Studies* 2(2), 156–167.

Pallant, Julie. 2010. *SPSS Survival Manual.*

Pasaribu, Truly Almendo. 2020. "Challenging EFL students to read: Digital reader response tasks to foster learner autonomy." *Teaching English with Technology* 20(2):21–41.

Reinders, Hayo. 2010. "Encouraging autonomy with an online language support system." *CALL-EJ Online* 11(2):1–16.

Reinders, Hayo. 2018. "Research agenda: Language learning beyond the classroom". *Language Teaching* 50:561–578.

Reinders, Hayo, and Cynthia White. 2016. "20 years of autonomy and technology: How far have we come and where to next ?" *Language, Learning and Technology* 20(2):143–54.

Richards, Jack C. 2015. "The Changing Face of Language Learning : Learning Beyond the Classroom."

Richards, Jack C. 2015. "The changing face of language learning: learning beyond the classroom." *RELC Journal* 46(1):5–22.

Shams, Idée Edalati. (2013). "Hybrid learning and Iranian EFL learners' autonomy in vocabulary learning." *Procedia-Social and Behavioral Sciences* 93, 1587–1592.

Smith, Bryan. 2003. "Computer-Mediated Negotiated Interaction : An Expanded Model."
Tafazoli, Dara, Elena Gómez Parra, and Cristina A. Huertas Abril. 2017. "Computer literacy: Sine qua non for digital age of language learning & teaching." *Theory and Practice in Language Studies* 7(9):716–22.
Tassinari, Maria Giovanna, and Freie Universität Berlin. 2012. "Evaluating learner autonomy: A dynamic model with descriptors." *Studies in Self-Access Learning Journal* 3:24–40.
Thaheem, Sahib Khatoon, Mohamad Jafre Zainol Abidin, Quratulain Mirza, and Habib Ullah Pathan. 2021. "Online teaching benefits and challenges during pandemic COVID-19: A comparative study of Pakistan and Indonesia." *Asian Education and Development Studies.* doi:10.1108/AEDS-08-2020-0189.
Thorne, Steven L. 2014. The 'Intercultural Turn' and language learning in the crucible of new media. In F. Helm & S. Guth (Eds.), *Telecollaboration 2. 0 for Language and Intercultural Learning.* Bern: Peter Lang.
Toffoli, Denyze, and Geoff Sockett. 2015. "University teachers' perceptions of online informal learning of english (OILE)." *Computer Assisted Language Learning* 28(1):7–21.
Van Dozer, Carol. 1997. "Improving ESL Learners' Listening Skills : At the Workplace and Beyond." *Knowledge Creation Diffusion Utilization* (February):1–6.
Wang, Shiang-Kwei, and Hui-Yin Hsua. 2008. "Reflections on using blogs to expand in-class discussion." *TechTrends* 52:81–82.
Ward, Jason M. 2004. "Blog assisted language learning (BALL): Push button publishing for the pupils." *TEFL Web Journal* 3(1):1–16.
Warschauer, Mark. 1996. "Comparing face-to-face and electronic discussion in the second language classroom." *CALICO Journal* 13(2): 7–26.
Wenden, Anita L. 1997. "Autonomy in language learning." *System* 25(4):584–88.
Wilson, Amanda. 2018. "Book Review The 'Bigger Story': Reflections on Language Teacher Identity Research.".

6 The Role of Digital Innovation in E-Learning System for Higher Education during COVID 19

A New Insight from Pedagogical Digital Competence

Waqas Khan, Qasim Ali Nisar, and Saira Sohail
The Superior University Lahore

Sidra Shehzadi
Universiti Utara Malaysia

CONTENTS

DOI: 10.1201/9781003143796-6

6.1 INTRODUCTION

The year 2020 saw the spread of COVID-19 that led to the WHO declaring it a pandemic (CDC COVID-19 Response Team and McNamara, 2020), and this markedly affected every field including tutoring, administration, and community life as a result of precautionary measures adopted by the countries in the form of social distance (Shahzad et al., 2020) to stop the spread of the disease. The citizens of developing counties are also facing the same epidemic (Akram, Anjum, & Batool, 2020). Pakistan is tagged in the list of those developing countries that are struggling to push up their educational profile. Pakistan's social and living standards measurement survey 2018–2019 reported a 2% increase in literacy rate to 60% as compared to 58% according to last survey in 2015–2016 (Psalm, Government of Pakistan. June 2020). Another report published by the world economic forum indicated Pakistan at 94th place in quality education worldwide (www.pakistantoday.com.pk, 2018). The need for learning is evident from the literature, especially in developing countries (Ashraf & Hafiza, 2016). The current crisis has obstructed the way of learning and compels educational institutes to adapt to this new normal to handle the panic situation. Setiawan (2020) specified that 80% of pupils have been severely affected due to restricted movement during the pandemic. Developed countries are old beneficiaries of cybernetic learning environment, but higher education institutions in developing countries are trying to cope with the strain of digitalized learning culture to reduce educational dropout (Shahzad et al., 2020).

The second wave of the outbreak COVID-19 has prolonged the transformational learning environment. Hence, this modified method is not new in the world, but for developing countries, it is challenging due to restricted resources (Hasan, Rehman, & Zhang, 2021). UNESCO preferred virtual study as an alternative during COVID-19 (Bansal, 2020) and this was globally acknowledged without any country having any choice in the matter. Switching towards the digital world led educational organizations to utilize resources like the internet and related supporting gadgets (Radha, Mahalakshmi, Kumar, & Saravanakumar, 2020). Ocaña-Fernández, Valenzuela-Fernández, and Garro-Aburto (2019) emphasized that educational associations have made technological innovations by introducing different innovative instruments, digital skills, and artificial intelligence in their migrated system. Current e-learning research studies flesh up technology enhancement by adding the concepts of IoT (Internet of Things), Pedagogical digital competence, and

ICT adoption (information & communication technologies) (Foulger, Graziano, Schmidt-Crawford, & Slykhuis, 2017; Patel & Cassou, 2015; Ab Wahab, Mohamad, Yusuff, & Musa, 2020). All the above-mentioned resources are toolkits to fit the technological vehicle without any hindrance. As mentioned by Rodrigues, Almeida, Figueiredo, and Lopes (2019), online knowledge creation involves a set of tools and techniques provided to students to enable them to avail user-friendly, customized, and progressive learning abilities.

The problem occurring within the digital education mechanism (Tran et al., 2020) is still a hot topic of research. Recently, scholars are trying to find out new themes of e-learning through which stakeholders can get maximum benefits. Developing countries have a plethora of problems due to lack of infrastructure; especially in rural areas, students are facing a lack of availability of internet facilities. According to the results of a recent study, 26.86% of students have experienced technical issues and 19.36% have worse internet connections (Radha et al., 2020). Around 50% of scholars and facilitators have trouble with connectivity (Tran et al., 2020). Therefore, it is an important requirement of HEI's and governing bodies to furnish a protected and trouble-free digitalized learning environment. Past studies only begin to emphasize on continuity of virtual education (Ali, 2020). Lack of research in this regard has been found in literature in struggling countries. Especially in rising realms, no descriptive study has been found in this consideration. There always remains an insufficiency of knowledge as evolving trends mold the phenomena in new directions and researchers always try to cover new themes to align with emerging dimensions. Gupta and Gupta (2020) revealed the reality that the requirement of the new setup is changing in the technical structure of online lessons for stakeholders at all levels and it is a new task for research scholars. Till now, presented studies covered online learning environments with limited attributes. Most of the academic work remains restricted to the experimental type (Shahzad et al., 2020), review of literature studies (Fernández-Batanero, Román-Graván, Reyes-Rebollo, & Montenegro-Rueda, 2021; Rodrigues et al., 2019), comparison evaluation (Shafique, Khawaja, Sabir, Qazi, & Mustaqim, 2020), some meta assessments (Ali, 2020), and commentary reviews (Tran et al., 2020).

Statistical analysis in this field of innovative environment is still missing. Previously, researchers tried to analyze artificial intelligence, social IoT, and ICT adoption, as isolated phenomena worldwide mostly in advanced countries; hence, the collaborated studies which tied the above stated concepts in one knot are missing. Especially, at the global level, this study is going to be the first step in this respect. Pedagogical digital competence and computer self-efficacy have not been studied at the individual level or as a mediator in developing country's virtual learning themes on a single page. This academic writing is rare participation in knowledge enhancement by examining PDC and computer self-efficacy as mediators, act as facilitators in the migrating system. Technostress (Fernández-Batanero et al., 2021) and Techno overload (Kamal, Rabbani, Samdani, Shujaat, & Ahmad, 2020) are two extremely important factors when dealing with high technology. These contents have been studied in the social (Media) therefore, at the level of education, nowhere is it to be found in the context of underdeveloping boundaries. Educational norms have been changed due to unexpected circumstances and online schooling is dissimilar from routine scholarship (Shahzad et al., 2020). There is a need not only to recognize the required system but also to realize the innovative theme to continue to update the system for the best results. This academic writing is a tiny effort in striving to link the gap that has not been disclosed earlier. The framework presented in this fresh book chapter involves artificial intelligence, social IOT, and ICT adoption as independent variables impacting the e-learning system in the presence of mediators, pedagogical digital competence, and computer self-efficiacy, along with the of moderators technostress and techno overload, in developing countries.

Understanding and implementing the new approaches of innovative mechanism are prerequisites for new emergency trends. Hence, current research work is going to provide a significant input in literature of online learning.

6.2 LITERATURE REVIEW

The below-mentioned aspects highlight the results of earlier studies for theoretical support.

6.2.1 Hypothesis Development

6.2.1.1 ICT Adoption, Pedagogical Digital Competence, Computer Self-Efficacy, and E-Learning System

Mirete, Maquilón, Mirete, and Rodríguez (2020) endorsed that ICT is very important to enhance the pedagogical digital competence and for the success of the e-learning system. Laurillard (2007) expressed the relationship between ICT and pedagogical digital competence along with the efficiency of online teaching and e-learning. Ali (2020) highlighted ICT as a necessity for the higher education institutes with the functionality of digital skills and confidence, especially during COVID-19. Moreover, Ali, Uppal, and Gulliver (2018) described the relationship and well-built relationship of ICT with student effectiveness during e-learning system. Another study highlighted the importance of ICT along with e-learning to improve the higher education institute quality and also improve the digital competence and skills of teachers and student effectiveness (Pavel, Fruth, & Neacsu, 2015). Yet another study stated that ICT adoption is associated with computer self-efficacy and ICT self-efficacy, especially in an e-learning system (Hatlevik & Hatlevik, 2018). Dong, Xu, Chai, and Zhai (2019) explored the structural relationship of pedagogical digital competence and computer self-efficacy along with an e-learning system. The study results revealed that ICT adoption enhances the teacher's digital competence, skills, and self-efficacy by use of computers with the e-learning system (Brush, Glazewski, & Hew, 2008; Graham, Borup, & Smith, 2012; Shehzadi et al., 2020). Boelens, De Wever, and Voet (2017) and Drent and Meelissen (2008) also endorsed that integration of ICT is enhancing the teacher's pedagogical competence, skills, and computer self-efficacy. Furthermore, by using ICT in teaching to improve the pedagogical digital competence, computer self-efficacy, and e-learning system, success can be achieved (López-Vargas, Duarte-Suárez, & Ibáñez-Ibáñez, 2017; Semiz & Ince, 2012). Keeping in view above, this study hypothesized that:

H1: There is a significant relationship between ICT adoption and pedagogical digital competence.

H2: There is a significant relationship between ICT adoption and computer self-efficacy.

H3: There is a significant relationship between ICT adoption and e-learning system.

6.2.2 Social IoT, Pedagogical Digital Competence, Computer Self-Efficacy, and E-Learning System

Sánchez, Mallorquí, Briones, Zaballos, and Corral (2020) highlighted that social IoT is worked through pedagogical digital strategy and enables the ability for online teaching and e-learning system. Tűrkeş et al. (2020) endorsed that the use of Social IoT along with professional skills and digital competence provides effectiveness in online activity as well as enhances self-efficacy. The study results revealed that IoT and digital competence or skills and technology flexibility enhance the demand of end-users (student and teachers) in the digitalization era (Duboc, Kinnear, Alsina-Pagès, & Alías, 2019). Moreover, higher education institutes in Europe run all the undergraduate and postgraduate programs through e-learning system by use of IoT functionality and digital competence that acts as a pillar between teachers and students (Grossman, Reid, & Morgan, 2001; Levie, 2014; Stentoft, 2017). Rushby and Surry (2016) highlighted the importance of the use of technology for e-learning system and also endorsed the role of digital strategies (pedagogical digital competence). Majeed and Ali (2018) elaborated the relationship of IoT and learning system with remarkable development in real life as well as highlighting its ease. Eastin and LaRose (2000) described the relationship between computer self-efficacy and the role of technology in the completion of the real-life tasks (e-learning). Hsu and Lin (2018) described that IoT brings together the different devices and provides facilitation through digital competence and computer self-efficacy, but has several limitations in the form of securities issues. The study results thus prove that a positive attitude exists regarding social IoT and learning system or other fields of work including the competence of digitalization with quality services (Kim, 2016; Liu, Chen, Solanas, & He, 2017). As per this scenario, the current study hypothesized that:

H4: There is a significant relationship between social IoT and pedagogical digital competence.

H5: There is a significant relationship between social IoT and computer self-efficacy.

H6: There is a significant relationship between social IoT and e-learning system.

6.2.3 Artificial Intelligence, Pedagogical Digital Competence, Computer Self-Efficacy and E-Learning System

Ocaña-Fernández et al. (2019) elaborated that the challenge of a new millennium was for universities to develop the technological environment in an e-learning system by using artificial intelligence along with pedagogical digital competence or computer skills. Chatterjee and Bhattacharjee (2020) highlighted the main theme of the use of artificial intelligence to enhance the digital competence or computer efficacy of teachers and students and the provision of satisfaction during e-learning. Furthermore, Kumar (2019) endorsed the relationship of artificial intelligence and online learning with other fields of life. Ahmad (2019) also described the importance of artificial intelligence in the context of facilitation of higher education e-learning along with digital competence and skills. Fahimirad and Kotamjani (2018) stated the role of artificial intelligence to improve the quality of educational institutes during e-learning that boosts the teacher's digital skills in addition to challenges of e-learning. The study results exposed the innovation in technology by use of artificial intelligence in e-learning enhance the digital skills of students and teachers in competitive environment (Kena et al., 2014; Phillips, 2012). Kena et al. (2014) expressed that the interface of artificial intelligence and computer self-efficacy is drawing attention to new approaches in the digitalization era with the functionality of digital skills. The studies also highlighted the relationship between artificial intelligence and the e-learning educational system and various types of information and communication challenges (Ferguson et al., 2015; Perez, Deligianni, Ravi, & Yang, 2018). Ocaña-Fernández et al. (2019) described the most considerable challenge in e-learning as being able to integrate the technology with AI and enable adoption. In light of above, we research hypothesized that:

H7: There is a significant relationship between AI and pedagogical digital competence.
H8: There is a significant relationship between AI and computer self-efficacy.
H9: There is a significant relationship between AI and e-learning system.

In the light of the above-said debate of existing studies, the research highlighted that ICT, social IoT, and artificial intelligence work as key technology drivers in technology adoption for e-learning system and that pedagogical digital competence and computer self-efficacy work as key supporting elements. All these constructs are needed in the technology driver's tool kit. Keeping in mind these points, we study hypothesized that:

H10: There is a significant relationship between pedagogical digital competence and e-Learning system
H11: There is a significant relationship between computer self-efficacy and e-learning system
H12: Pedagogical digital competence mediates the relationship between ICT adoption and e-learning system
H13: Pedagogical digital competence mediates the relationship between social IoT and e-learning system
H14: Pedagogical digital competence mediates the relationship between AI and e-learning system
H15: Computer self-efficacy mediates the relationship between ICT adoption and e-learning system
H16: Computer self-efficacy mediates the relationship between social IoT and e-learning system

H17: Computer self-efficacy mediates the relationship between AI and e-learning system

6.2.4 Moderating Role of Technostress

Joo, Lim, and Kim (2016) indicated that pedagogical digital competence is developing a significant relationship with technostress and e-learning system as well as enhancing the teacher's efficiency in the context of the use of technology. The results also revealed that technostress performed a role as a moderator. Yin, Lim, Sun, and Zhao (2012) endorsed that technostress intensity increased after the adoption of technology in South Korea and it will be controllable with digital competence; otherwise, technostress will further worsen. The term technostress was first introduced by Weil and Rosen (1997) who elaborated on the negative relationship of technostress with other fields of work such as e-learning due to the adoption of technology. Similarly, Wang, Shu, and Tu (2008) highlighted the same scenario of technostress by use of technology in real life. Chatzoglou, Sarigiannidis, Vraimaki, and Diamantidis (2009) and Rezaei, Mohammadi, Asadi, and Kalantary (2008) revealed that technostress is shown to have a negative impact on technology in the context of the e-learning system. Moreover, Jena (2015) described the role of technostress between the Indian academicians and the e-learning system environment in the context of digital competence. Effiyanti and Sagala (2018) recommended that technostress in e-learning will be reduced by the use of digital competence with computer self-efficacy, and the results are based on the performance of teachers in online learning. Besides, Bradley and Russell (1997) also suggested that the reduction of teachers' technostress in e-learning system to enhance the use of computer self-efficacy and digital competence. Kay (2008) stated that digital skills, training, computer efficacy, and teacher's knowledge reduced the technostress of teachers on the e-learning platform. In the 21st century, the use of ICTs is very suitable for every individual in a collaborative learning environment, but technology adoption also created obstacles in the context of technostress that should be overcome by the use of computer and digitalization skills (Ayyagari, 2012; Ayyagari, Grover, & Purvis, 2011). In the light of the above-said points, we hypothesized that:

H18: Technostress moderates the relationship between computer self-efficacy and e-learning system
H19: Technostress moderates the relationship between pedagogical digital competence and e-learning system

6.2.5 Moderating Role of Techno overload

Karr-Wisniewski and Lu (2010) endorsed the relationship of techno overload and digital competence or skills. The results highlighted that techno overload should be manageable by the use of digital competence and self-efficacy skills. Moreover, Al-Abdullatif, Alsubaie, and Aldoughan (2020) elaborated that mobile technostress in the context of techno overload has shown a significant impact on students' learning and that pedagogical digital or technical competence will overcome the impact of technology overload. Whyte (2014) analyzed the pedagogical digital skills and found that self-efficacy reduced the techno overload impression during online teaching by using social media development regarding techno pedagogical skills. The study results explained that techno-overload factors have shown a negative impact on e-learning system especially during COVID-19 (Christian, Purwanto, & Wibowo, 2020). In addition, Li and Wang (2020) demonstrated the relationship between techno overload and university teachers' performance. The results displayed that technology overload creates various troubles for teachers during the e-learning system and results in the ineffectiveness of students in the e-learning system. Techno overload is created when the teachers encourage faster as well as longer use of technology, and its results is obstacles in online learning (Funminiyi, Akinlolu, & Agboola, 2014). Werang (2018) explained the positive relationship between techno workload and teacher's emotions in the context of teacher's e-learning efficiency. Ibrahim and Yusoff (2015) described the relationship between techno overload and government agencies in Malaysia. The results also endorsed that digital competence skills and computer self-efficacy overcome the techno overload after effects. Hence, we hypothesized that:

H20: Techno overload moderates the relationship between computer self-efficacy and e-learning system

H21: Techno overload moderates the relationship between pedagogical digital competence and e-Learning system

6.3 THEORETICAL FOUNDATION

6.3.1 Theory of Reasoned Action (TRA) Model

The Theory of Reasoned Action (TRA) model also referred to as a theory of reasoned action was first introduced by Ajzen and Fishbein (1988). The TRA model highlights the individual's behavior intention to perform a specific task and, further, this intention is to turn into behavior norms as well as subjective norms regarding execution or completion of the task. The theory consists of a four-step procedure.

- The first step consists of the behavioral beliefs and evaluation (BBE) and normative beliefs and motivation to comply (NMBC). The BBE is represented as the belief of an individual to perform a specific task that will help and be evaluated after implementation. The NMBC represents an individual normative belief to perform a specific task by complying with the motivational factors.
- The second stage describes attitude towards the behavior (ATB) and the subjective norm (SN). ATB represents the degree after executing a task with favorable or unfavorable appraisal or evaluation of behavior regarding the task. SN highlights the social pressure to perform or not perform a task with the functionality of behavior.
- In the third stage, behavior intention (BI) is shown and it is described as the degree to which an individual observes the conscious plan to perform or not perform the task due to external pressure.

- In the last and fourth stage, the actual behavior (AB) is executed and the task is completed.

The theory of reasoned action was also amended in the context of future behavioral task by Warshaw and Davis (1985). Sheppard, Hartwick, and Warshaw (1988) also accomplished a meta-analysis on the helpfulness of the TRA model in real life. Besides Otieno, Liyala, Odongo, and Abeka (2016) conducted a review in the context of a theoretical comparison of the TRA model with other theories such as TAM, TOE, TPB, and UTAUT. The current study is also supported and parallels of TRA model. In this book chapter, the ICT adoption is referred to as BBE. The Social IoT and Artificial Intelligence is executed as an NMBC. In second stage, the pedagogical digital competence and computer self-efficacy perform as ATB and SN. In the third step, the technostress and techno overload act as a conscious plan in shape of BI. In the last and fourth step, AB is performed in context of e-learning system.

The chapter has elaborated the role of ICT adoption, Social IoT, and artificial intelligence for the implementation of e-learning system during COVID-19. The pedagogical digital competence and computer self-efficacy are also used as supporting key elements for the successful implementation of e-learning system during COVID-19. The study also elaborated the factors that affected the e-learning system during implementation, such as technostress and techno overload. In the light of the above-said debate of existing studies and theories, we propose a model (Figure 6.1):

6.4 METHODOLOGY

6.4.1 Sample and Data Collection

The current study used the quantitative approach to highlight the role of key technology drivers in the e-learning system (Merriam & Grenier, 2019) and test the

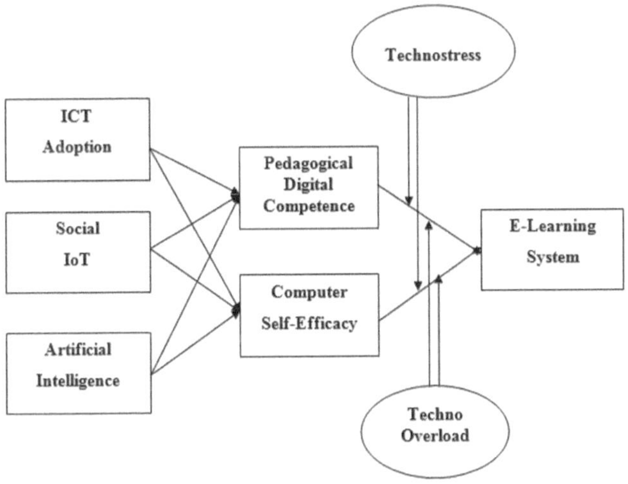

FIGURE 6.1 Visual image of the research

hypothesis by use of descriptive research structure (Dulock, 1993). With the survey questionnaires that were prepared by use of a Likert scale, the primary data were gathered for this study (Hox & Boeije, 2005). The simple random sampling technique is used for data collection due to the fact that this type of sampling provides the full opportunity to the population with an equal selection of the choices (Mehmood, Liland, Snipen, & Sæbø, 2012). The target population included the directorate of information technology (DIT) of higher education institutes (HEIs) of Pakistan. A total of 230 higher education institutes and 113 sub-campuses of those institutes are accredited by the Higher Education Commission of Pakistan (HEC, Islamabad; www.hec.org.pk). The survey questionnaires were sent through email to each higher education institute DIT. The total number of responses received from HEIs was 271, and the response rate was 78.71%. The 259 survey questionnaires were accepted for empirical analysis and the remaining 12 were rejected due to missing values.

6.4.2 Measures

For this chapter, the measures were adopted from existing studies. In ICT adoption, seven measures are used for that and are adapted from Ruiz, Sánchez, and Pina (2015). Furthermore, the social IoT consisted of eight measures adapted from Tűrkeş et al. (2020). In the case of artificial intelligence, four measures were used that were adapted from the study by Chatterjee and Bhattacharjee (2020). Additionally, a computer self-efficacy adapter from the study by Dong, Xu, Chai, & Zhai (2019) used five measures. Also, pedagogical digital competence consisted of eight measures through adoption (Mishra & Koehler, 2006). Moreover, technostress described 14 measures and was adapted from Lam, Cheng, and Choy (2010). The techno overload was adapted from Karr-Wisniewski and Lu (2010) by use of ten measures.

Likewise, the e-learning system used the five measures that were adapted from Jena (2015). Table 6.1 briefly provides the detail of measures with adapter references.

6.5 EMPIRICAL RESULTS

The study used PLS-SEM for empirical analysis that is a modernized measurement instrument in the current decade (Ali, Rasoolimanesh, Sarstedt, Ringle, & Ryu, 2018). PLS-SEM is also considered an elastic analysis tool (Ringle, Wende, & Will, 2005). Joseph F Hair Jr, Hult, Ringle, and Sarstedt (2016) elaborated that PLS-SEM is incredibly appropriate due to smaller data conditions with data normality. The above-said characteristics of the study suggested a Smart PLS-3 for the analysis and estimation of the hypothesis. The study employed the two-step process for presentation and calculation of results (Henseler, Ringle, & Sinkovics, 2009). This two-step process is the most suitable method in the social science research work field (Henseler et al., 2009). The first step denotes the measurement model assessment analysis through Confirmatory Factor Analysis (CFA) by using the consequences of convergent as well as discriminant validity. The second step employs the structural model assessment (SEM) by using the consequences of path analysis, interaction term, and indirect effect.

TABLE 6.1
Variables Measures

Variables	Items	Measures	Adaption from Studies
Information & Communication Technology (ICT) Adoption	07	ICTs in teaching, ICTs in e-class rooms ICTs in the e-learning system	Ruiz et al. (2015)
Social Internet of Things (IoT)	08	IoT understanding, IoT & control of ecosystem, IoT challenges,	Tűrkeş et al. (2020)
Artificial Intelligence (AI)	04	AI in HEIs, AI in the teaching, AI in the e-class rooms, AI in the society	Chatterjee and Bhattacharjee (2020)
Computer Self-Efficacy (CSE)	05	Technical skills Confidently solve problems by use of ICT	Dong et al. (2019)
Pedagogical Digital Competence(PDC)	08	Online digitalization environmental competence, Modest online interactivity, Encouragement of online interactivity, Creative effective representations	Mishra and Koehler (2006)
Technostress (TS)	14	Techno-complexity, Techno-insecurity, Techno-uncertainty	Lam et al. (2010)
Techno Overload	10	Information-overload, System feature-overload, Communication- overload	Karr-Wisniewski and Lu (2010)
E-Learning System (ELS)	05	A collaborative environment for teaching & research, a Collaborative environment for students & colleagues, Professionalism in the teaching task	Jena (2015)

Source: Author's Design.

6.5.1 Data Normality

PLS-SEM is a nonparametric method for inferential statistics and as per threshold, no data will be run for inferential analysis without the confirmation of data normality (Hair, Money, Samouel, & Page, 2007). In this context, the study recommends to run the normality distribution test to check the data's normal position. The study used the histogram plot method, skewness, as well as kurtosis techniques for data normality distribution (Munro, 2005). The value of the data normality test falls between −2 and +2. Meanwhile, results elaborated that results fall under threshold and data is normally distributed. Therefore, the study used further estimation techniques for analysis.

6.5.2 Common Method Biases

In this book chapter, the samples for data collection were gathered from higher education institute DITs; hence, the data could have common method bias issues. In the light of the above, the multicollinearity test is compulsory to make sure the issue of common method biases in data Kock (2015) is not there. The value of the collinearity test falls within the boundary of 3.3. Meanwhile, results elaborated that result of all variables fall under the boundary of 3.3, highlighting that data was no issue in the context of CMB. Therefore, the study used further estimation techniques for analysis through PLS-SEM.

6.5.3 Measurement Model Assessment

In the first stage, the measurement model assessment technique is used through the functionality of CFA with supportive tests of convergent validity and discriminant validity. Meanwhile, convergent and discriminant validity tests explained the data reliability and validity in the context of an internal item of constructs. Therefore, convergent validity elaborated the CR, AVE, Alpha value, and factor loadings. Table 6.2 and Figure 6.2 endorse that mostly loading values of constructs are greater than 0.70 and their indication of the significance of factor loading as per threshold value (Hair Jr., Sarstedt, Ringle, & Gudergan, 2017). The data consistency is also highlighted through composite reliability (CR). Table 6.2 authorizes that CR values of constructs are greater than 0.70 and its indication of the significance of CR as per threshold value is more than 0.70 (Joe F Hair Jr, Sarstedt, Hopkins, & Kuppelwieser, 2014). The data reliability is also checked by the use of the Alpha value. Table 6.2 certified that Alpha values of constructs are also greater than 0.70, and its indication of the significance of Alpha test as per threshold value that is 0.70 (Hair Jr. et al., 2014). Moreover, Average variance extracted (AVE) described the data internal validity of each construct. Table 6.2 sanctions that AVE values of constructs are greater than 0.50 and its indication of the significance of data internal validity as per threshold value is 0.50 (Cheung & Wang, 2017). Besides, the validity of data is also checked by the use of discriminant validity tests through Hetreotrait-Monotrait Criteria (HTMT ratio). Table 6.3 expressed that every variable HTMT ratio is less than 0.85 and its indication of the significance of discriminant validity as per threshold value that is 0.85 (Kline et al., 2012).

6.5.4 Structural Model Assessment

The second stage expressed the structural model assessment through the functionality of Structure equation model (SEM) by use of path analysis, interaction term, and indirect effect with the help of bootstrapping process (Ringle, Wende, & Will, 2005). Meanwhile, hypotheses are verified to check the impact of key technology drives tool kit on e-learning system during COVID-19. In the below-mentioned table's beta values, p values, t values as well as standard error results expressed the authentication of hypothesis accepting or rejecting and model confirmation in appearance of significant and insignificant. Table 6.4 and Figure 6.3 demonstrate that information and communication technology (ICT) has developed a significant relationship with pedagogical digital competence as well as e-learning system with the following values: $\beta=0.359$, $t=4.850$, $LL=0.230$, $UL=0.473$; $\beta=0.143$, $t=2.163$, $LL=0.043$, $UL=0.250$. Hence, H1 and H3 are significant and empirically proofed. Meanwhile,

TABLE 6.2
Convergent Validity (Confirmatory Factor Analysis)

Constructs	Items	Loading	Alpha	CR	AVE
ICT Adoption	ICT1	0.732	0.887	0.911	0.596
	ICT2	0.740			
	ICT3	0.750			
	ICT4	0.766			
	ICT5	0.825			
	ICT6	0.814			
	ICT7	0.771			
Social IoT	IoT2	0.672	0.834	0.875	0.503
	IoT3	0.585			
	IoT4	0.691			
	IoT5	0.768			
	IoT6	0.812			
	IoT7	0.690			
	IoT8	0.727			
Artificial Intelligence	AI1	0.827	0.836	0.891	0.674
	AI2	0.888			
	AI3	0.708			
	AI4	0.849			
Pedagogical Digital Competence	PDC1	0.784	0.911	0.928	0.616
	PDC2	0.769			
	PDC3	0.823			
	PDC4	0.792			
	PDC5	0.786			
	PDC6	0.771			
	PDC7	0.766			
	PDC8	0.790			
Computer Self-Efficacy	CSE1	0.751	0.854	0.894	0.629
	CSE2	0.822			
	CSE3	0.896			
	CSE4	0.683			
	CSE5	0.798			
Technostress	TS11	0.647	0.889	0.906	0.619
	TS12	0.848			
	TS13	0.813			
	TS14	0.696			
	TS2	0.801			
	TS5	0.890			
Techno overload	TOL1	0.596	0.773	0.832	0.500
	TOL10	0.716			
	TOL4	0.677			
	TOL6	0.707			
	TOL7	0.819			

(Continued)

TABLE 6.2 (*continued*)
Convergent Validity (Confirmatory Factor Analysis)

Constructs	Items	Loading	Alpha	CR	AVE
E-Learning System	ELS1	0.811	0.867	0.904	0.653
	ELS2	0.819			
	ELS3	0.833			
	ELS4	0.821			
	ELS5	0.754			

TABLE 6.3
Discriminant Validity (HTMT Ratio)

Constructs	AI	CSE	ELS	ICT	PDC	Social IoT	TOL	TS
AI								
CSE	0.116							
ELS	0.625	0.162						
ICT	0.526	0.089	0.566					
PDC	0.585	0.127	0.717	0.605				
Social IoT	0.449	0.080	0.460	0.538	0.460			
TOL	0.119	0.176	0.124	0.120	0.089	0.116		
TS	0.064	0.063	0.090	0.155	0.101	0.104	0.080	

Note: ICT, Information & communication technology adoption; Social IoT, Social internet of things; AI, Artificial Intelligence; PDC, Pedagogical digital competence; CSE, Computer Self-Efficacy; TS, Techno stress; TOL, Techno overload; ELS, E-learning system.

ICT developed an insignificant relationship with compute self-efficacy ($\beta=0.023$, $t=0.298$, $LL=-0.099$, $UL=0.160$); therefore, H2 is not supported. Likewise, the social IoT (Internet of thing) is also described as having a significant relationship with pedagogical digital competence as well as e-learning system ($\beta=0.120$, $t=1.361$, $LL=0.005$, $UL=0.293$; $\beta=0.092$, $t=1.293$, $LL=0.006$, $UL=0.209$). Hence, H4 and H6 are significant and empirically verified. Meanwhile, the social IoT (Internet of thing) developed an insignificant relationship with computer self-efficacy ($\beta=-0.015$, $t=0.173$, $LL=-0.145$, $UL=0.126$); therefore, H5 is not supported. Moreover, artificial intelligence is having a significant impact on the pedagogical digital competence, computer self-efficacy, and e-learning system ($\beta=0.306$, $t=3.177$, $LL=0.133$, $UL=0.453$; $\beta=0.109$, $t=1.429$, $LL=0.020$, $UL=0.229$; $\beta=0.208$, $t=3.000$, $LL=0.080$, $UL=0.327$). However, H7, H8, and H9 are supported and empirically approved. Also, pedagogical digital competence and computer self-efficacy generate a significant impact on the e-learning system ($\beta=0.392$, $t=4.594$, $LL=0.235$, $UL=0.513$; $\beta=0.082$, $t=1.258$, $LL=0.013$, $UL=0.164$); thus, H10 and H11 are supported and empirically confirmed.

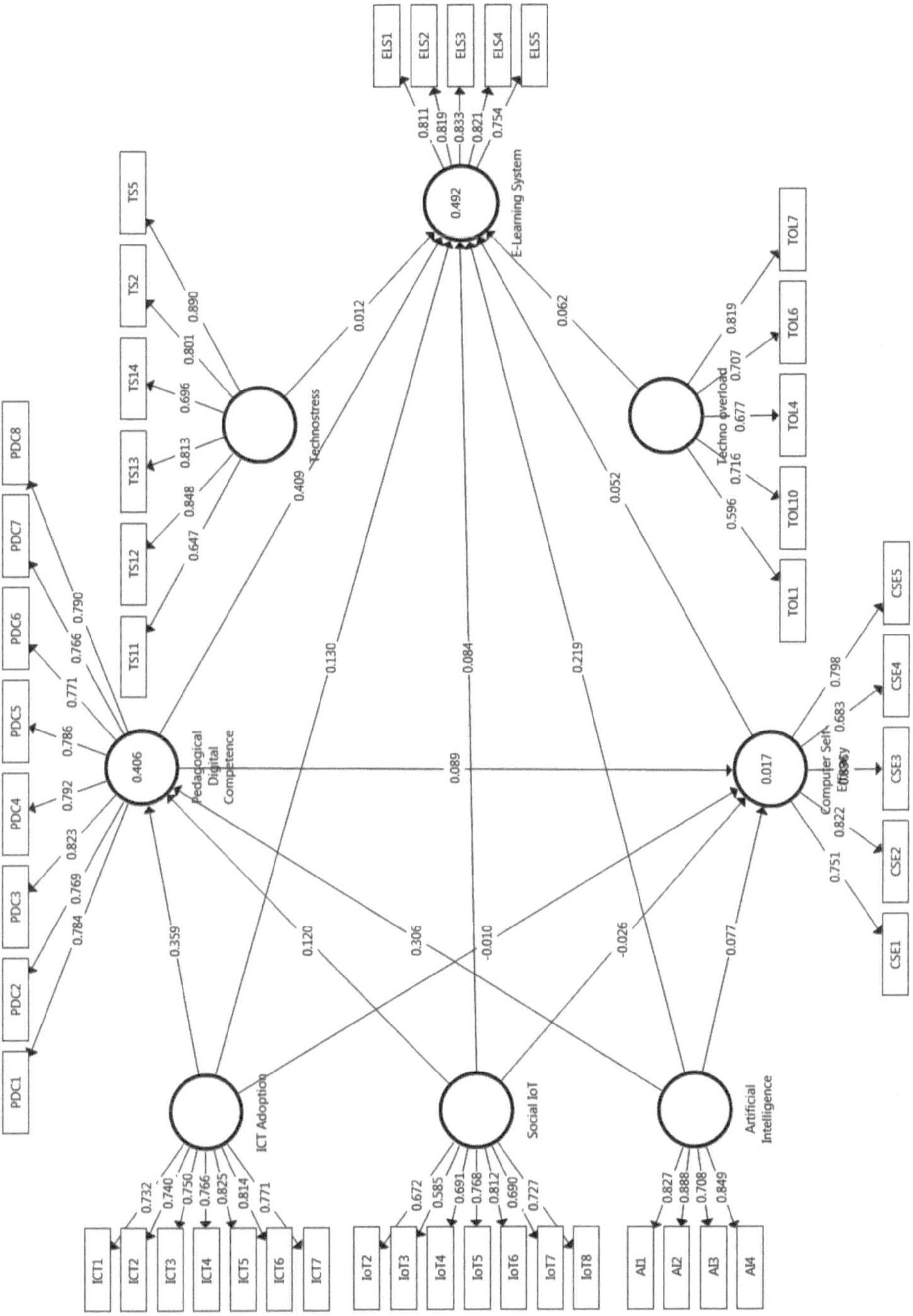

FIGURE 6.2 Measurement model assessment.

TABLE 6.4
Structural Equation Modeling (Path Analysis)

Hypothesis	Relationship	B	SD	T	p Values	LL	UL	Decision
H1	ICT -> PDC	0.359	0.072	4.850	0.000	0.230	0.473	Significant
H2	ICT -> CSE	0.023	0.078	0.298	0.383	−0.099	0.160	Insignificant
H3	ICT -> ELS	0.143	0.065	2.163	0.014	0.043	0.250	Significant
H4	IoT -> PDC	0.120	0.090	1.361	0.041	0.005	0.293	Significant
H5	IoT -> CSE	−0.015	0.084	0.173	0.431	−0.145	0.126	Insignificant
H6	IoT -> ELS	0.092	0.065	1.293	0.005	0.006	0.209	Significant
H7	AI -> PDC	0.306	0.097	3.177	0.001	0.133	0.453	Significant
H8	AI -> CSE	0.109	0.076	1.429	0.050	0.020	0.229	Significant
H9	AI -> ELS	0.208	0.075	3.000	0.003	0.080	0.327	Significant
H10	PDC -> ELS	0.392	0.084	4.594	0.000	0.235	0.513	Significant
H11	CSE -> ELS	0.082	0.046	1.258	0.038	0.013	0.164	Significant

ICT, Information & communication technology adoption; Social IoT, Social internet of things; AI, Artificial Intelligence; PDC, Pedagogical digital competence; CSE, Computer Self-Efficacy; ELS, E-learning system.

TABLE 6.5
Mediation Analysis (Indifferent Effect)

Hypothesis	Relationship	B	SD	T	*p* Values	LL	UL	Decision
H12	ICT -> PDC -> ELS	0.140	0.045	3.109	0.001	0.065	0.214	Significant
H13	IoT -> PDC -> ELS	0.047	0.034	1.399	0.035	0.002	0.109	Significant
H14	AI -> PDC -> ELS	0.120	0.043	2.806	0.003	0.044	0.186	Significant
H15	ICT -> CSE -> ELS	0.001	0.005	0.240	0.405	−0.008	0.010	Insignificant
H16	IoT -> CSE -> ELS	−0.001	0.007	0.123	0.451	−0.011	0.008	Insignificant
H17	AI -> CSE -> ELS	0.006	0.007	0.804	0.011	0.003	0.021	Significant

ICT, Information & communication technology adoption; Social IoT, Social internet of things; AI, Artificial Intelligence; PDC, Pedagogical digital competence; CSE, Computer Self-Efficacy; ELS, E-learning system.

With the parallel of direct relationship, the book chapter also demonstrates a mediator role of pedagogical digital competence and computer self-efficacy between ICT, Social IoT as well as AI. Table 6.5 and Figure 6.3 endorse that pedagogical digital competence is mediating significant relationship between ICT, social IoT, AI, and e-learning systems ($\beta=0.140$, $t=3.109$, $LL=0.065$, $UL=0.214$; $\beta=0.047$, $t=1.399$, $LL=0.002$, $UL=0.109$; $\beta=0.120$, $t=2.806$, $LL=0.044$, $UL=0.186$). Consequently, H12, H13, and H14 are empirically verified and supported. However, computer self-efficacy is not mediating a significant relationship among ICT, social IoT, and e-learning system ($\beta=0.001$, $t=0.240$, $LL=-0.008$, $UL=0.010$; $\beta=-0.001$,

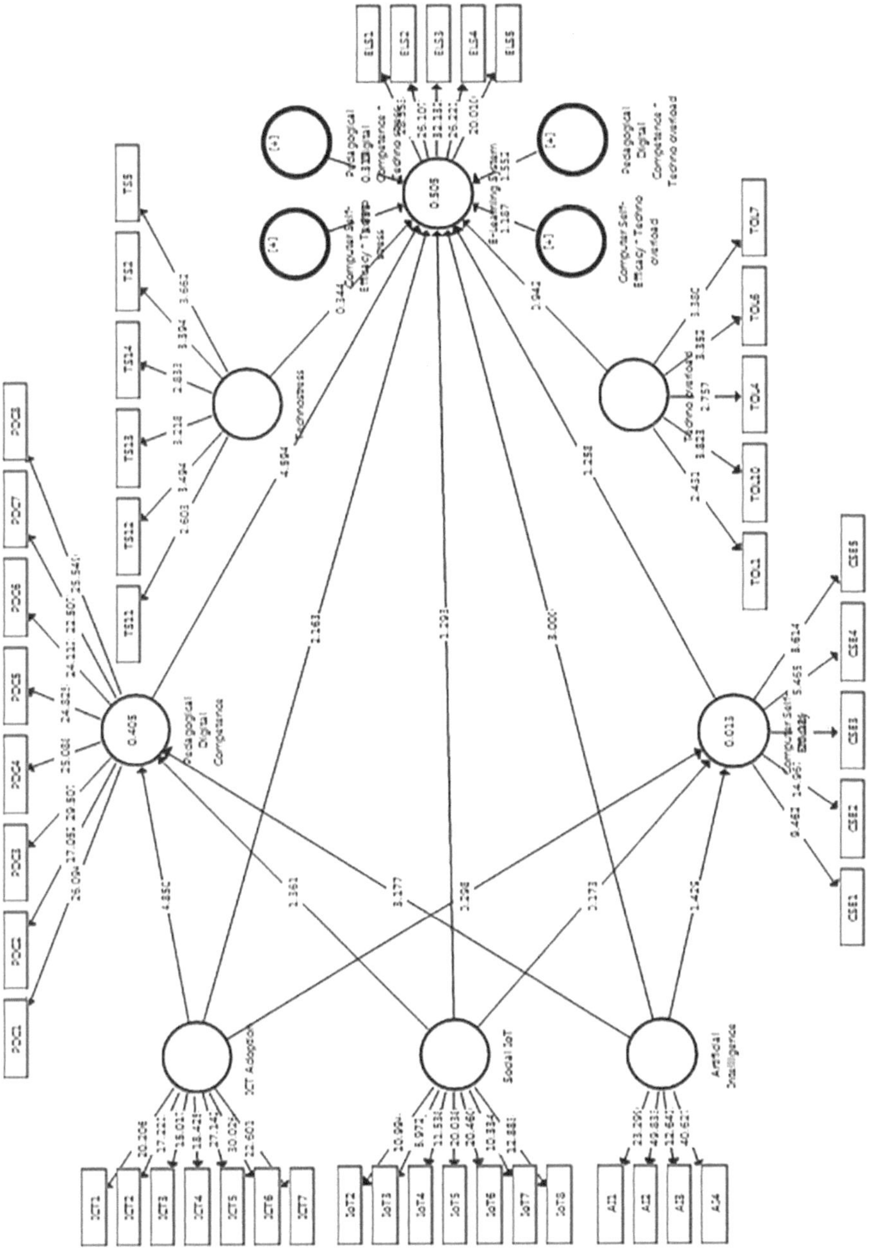

FIGURE 6.3. Structural model assessment.

TABLE 6.6
Moderation Analysis (Interaction Term)

Hypothesis	Relationship	B	SD	T	*p* Values	LL	UL	Decision
H18	PDC * TS -> ELS	0.018	0.056	0.319	0.375	−0.072	0.108	Insignificant
H19	CSE * TS-> ELS	0.077	0.050	1.559	0.050	0.024	0.138	Significant
H20	PDC * TOL -> ELS	−0.086	0.056	1.552	0.061	−0.151	0.031	Insignificant
H21	CSE * TOL -> ELS	0.050	0.042	1.187	0.118	−0.029	0.106	Insignificant

ICT, Information & communication technology adoption; Social IoT, Social internet of things; AI, Artificial Intelligence; PDC, Pedagogical digital competence; CSE, Computer Self-Efficacy; TS, Techno stress; TOL, Techno overload; ELS, E-learning system.

t=0.123, LL=−0.011, UL=0.008); accordingly, H15 and H16 are neither supported nor empirically approved. Therefore, computer self-efficacy is mediating a significant relationship among AI and e-learning system (β=0.006, t=0.804, LL=0.003, UL=0.021); accordingly, H17 is supported.

The current study is also describing a moderator role of technostress and techno overload between pedagogical digital competence (PDC), computer self-efficacy (CSE), and e-learning system. The Table 6.6 and Figure 6.3 highlight that technostress is not performing a moderator role among PDC and ELS (β=0.018, t=0.319, LL=−0.072, UL=0.108); therefore, H18 is not empirically approved and not supported. However, technostress is performing a moderator role among CSE and ELS (β=0.077, t=1.559, LL=0.024, UL=0.138); therefore, H19 is empirically approved and supported. Furthermore, techno overload is not executed as a moderator role among PDC, CSE, and ELS (β=−0.086, t=1.552, LL=−0.151, UL=0.031; β=0.050, t=1.187, LL=−0.029, UL=0.106); thus, H20 and H21 are not supported.

6.6 DISCUSSION AND CONCLUSION

Online courses have already remained a part of the academic system in western states like the USA where 89% of courses are through distance learning and full degree programs constitute for 50%. On the other hand, the picture is different in developing countries. In developing countries, only select universities have facilitated their pupils to get computer-based knowledge in very few courses due to lack of resources (Mumtaz, Saqulain, & Mumtaz, 2021). Hence, the COVID-19 spread did not prove to be awkward in the field of education for advanced nations but was troublesome for the underdeveloped world due to the pressure of quarantine. This compromising situation has poised the developing nations towards an immediate need of developing and applying a transformed educational system with consistent, economical, and protected tools (Mumtaz et al., 2021). Therefore, this study is a slim chance to significantly quench the thirst for knowledge by introducing enhanced contents in E-learning arrangement. According to the analysis upshots, introduction of advanced tools and techniques like ICT adoption, social IoT, and artificial intelligence significantly impact the E-learning environment in the presence of some mediating factors like pedagogical digital competence and computer self-efficacy along with moderating ingredients technostress and techno overload which have not been done before. Results elaborated that the impact of information and communication technologies on PDC and ELS is significant. This proved to be aligned with results in the literature (Ali, 2020; Boelens et al., 2017; Drent & Meelissen, 2008). While no significant link was found between ICT adoption and CSE, this means that the adoption of ICT can boost up this converted educational scheme (e-learning system) with the help of PDC. The hypothesis related to social IOT with PDC and ELS has been accepted and shows a significant association relatively as allied to literature (Abbasy & Quesada, 2017; Liu, Chen, Solanas, & He, 2017; Majeed & Ali, 2018; Sánchez et al., 2020; Tűrkeş et al., 2020).

However, the relationship between social IoT and CSE was determined insignificant. This reflected that despite limited studies, social IoT worked as part of an eco-system in online study mechanism for stakeholders to gets maximum benefits (Abbasy & Quesada, 2017). But still, in educational institutes, research is untouched and for maximum output and needs to furnish reliable investigation results. Statistical measurement yielded hypotheses 7–9, explaining the significant relationship between artificial intelligence with PDC, CSE, and ELS. Literature results have been verified here accordingly. As stated before, in HEIs, artificial intelligence is playing a momentous role. AI is a smart way to strengthen the PGC skills and digital abilities of all beneficiaries (Chatterjee & Bhattacharjee, 2020). During mediation analysis, PDC is significantly mediating between three independent variables and the dependent construct. So, the first three mediating hypotheses are accepted, whereas CSE only mediates between AI and ELS, and no mediation is found between ICT adoption and social IoT towards ELS. Mediation impact is partially within the relationships. Along with mediation, some other links measured are moderating factors. According to the calculations, moderation exists only within CSE and ELS by the moderating construct TS. The rest of the moderation-explaining hypotheses were rejected, means TS and TOL are insignificantly related to PDC and CSE.

The results depicted that to know more about the prevailing education system and technological advancement day by day, researchers need more ideas about the phenomenon. Studies till now are not enough to open new doors with new advancements. From the day of the pandemic, a digital system emerged with educational arrangements and a new form of technological tool has surprised us. The bond between man and machine has become stronger and enhancements take place all over the globe. This recent education system is the gift of a global pandemic in this regard. Many factors prevail in the society that directly or indirectly impact technology and environmental changes. These environmental changes create an opportunity for researchers, scholars, and all academicians to think about more suitable ways that facilitate stakeholders or users of technology as well as educationists at the global level. Against this background, this unique set of tools is the latest and slight contribution by researchers via taking individually studied technological phenomenon in a single drive. Researchers played a challenging role in this study by assessing single-studied moderating factors in a sole design and have discussed all possible technological requirements and problem creating variables in a solo strategy and presented a solution-oriented outcome. Excess of everything is bad and, accordingly, excessive use of technology is also dangerous but in the presence of digital skills, social IoT and ICT adoption on different levels make it easier, manageable, affordable, interactive, advantageous, and less stress in every regard. This is the requirement of time (new era), an efficient learning environment. Worldwide, HEI's are taking initiatives in this regard. Current research is the strategic solution for technological problems.

6.6.1 Theoretical Contribution

The current book chapter has produced a matchless contribution on the theoretical side in four components. First, the book chapter elaborated a unique theoretical concept in the current pandemic situation about the context of the adoption of the technology for e-learning system such as key technology drivers in name of "ICT adoption, social IoT, artificial intelligence, pedagogical digital competence, computer self-efficacy, technostress, techno overload". These drivers work as a key technology driver's tool kit in the e-learning system. Second, the book chapter endorsed the new arrival of the pedagogical digital competence and computer self-efficacy as a moderator that will enhance the efficiency of e-learning system especially in the time of COVID-19. Third, the current book chapter sanctioned the moderator role of technostress and techno overload that highlighted the psychological pressure of the use of technology especially in the era of COVID-19. Finally, the study also theoretically produced an unrivaled acquaintance with the theory of reasoned action, and this theory is most suitable and suggested on a priority basis in the context of the adoption of the technology and attitude towards the use of technology especially during COVID-19.

6.6.2 Practical Implications

In parallel to the theoretical manner, the book chapter also provides the practical side. The consequences of this research strongly suggested precious guidelines to the artificial intelligence designers, web developers, technology policymakers, information technology instructors, as well as researchers to deeply observe the main theme of distance learning or e-learning educational system, especially in the era of COVID-19 disease in the context of developing countries like Pakistan. In practical fieldwork, the current study has given unique directions to the university's higher authorities, higher degree-awarding institutions, education ministries, and management of directorate of information technology of universities for successful implementation of e-learning system along with the provision of teacher's competence and skills as well as after-effects of technological stress and overload. The current research is also very valuable for evaluation of teacher's performance or efficiency and student's satisfaction and effectiveness to use of technology for distance learning education system in the era COVID-19 disease in the context of developing countries like Pakistan. The current research is the solution of the next century digitalization era that specially focuses on the education system advancement worldwide.

6.6.3 Future Directions and Limitation

In parallel to the theoretical and practical manner, the current research has a few limitations. The current study purely focused on the distance learning education system in the context of technology adoption and student–teachers efficiency and effectiveness during COVID-19. In the light of the above, another study will be performed with a focus on other business activities. This fresh study is also purely spotlighting the implementation of technology for education in developing countries like Pakistan. In this scenario, the study will highlight only developing country experience. Thus, future studies will be performed by the focus on the implementation of technology for education in developed countries for comparative outcomes. In the end, the current study demonstrated the relationship of student teachers by use of technology during COVID-19. Thus, future studies will be performed focusing on the implementation of technology for organizations and professional businesses in developing and developed countries for comparative outcomes, especially in any crisis situation globally.

REFERENCES

Ab Wahab, N. Y., Mohamad, M., Yusuff, Y. Z., & Musa, R. (2020). The importance of ICT adoption in manufacturing sector: an empirical evidence on SME business performance. *International Journal of Supply Chain Management, 9*(2), 268–272.

Abbasy, M. B., & Quesada, E. V. (2017). Predictable influence of IoT (Internet of Things) in the higher education. *International Journal of Information Education Technology, 7*(12), 914–920.

Ahmad, T. (2019). Scenario based approach to re-imagining future of higher education which prepares students for the future of work. *Higher Education, Skills Work-Based Learning, 10*(1), 217–238.

Ajzen, I., & Fishbein, M. (1988). Theory of reasoned action-Theory of planned behavior. *University of South Florida, 2007*, 67–98.

Akram, M., Anjum, F., & Batool, Z. (2020). Covid-19: A reason behind digital education in Pakistan. *Mediterranean Journal of Social Sciences, 11*(4), 19–19.

Al-Abdullatif, A. M., Alsubaie, M. A., & Aldoughan, E. A. (2020). Exploring the effects of excessive texting through mobile applications on students' technostress and academic writing skills in the Arabic language. *IEEE Access, 8*, 166940–166950.

Ali, F., Rasoolimanesh, S. M., Sarstedt, M., Ringle, C. M., & Ryu, K. (2018). An assessment of the use of partial least squares structural equation modeling (PLS-SEM) in hospitality research. *International Journal of Contemporary Hospitality Management, 30*(1), 514–538.

Ali, S., Uppal, M. A., & Gulliver, S. R. (2018). A conceptual framework highlighting e-learning implementation barriers. *Information Technology People, 31*(1), 156–180.

Ali, W. (2020). Online and remote learning in higher education institutes: A necessity in light of COVID-19 pandemic. *Higher Education Studies, 10*(3), 16–25.

Ashraf, M. A., & Hafiza, I. (2016). Education and development of Pakistan: A study of current situation of education and literacy in Pakistan. *US-China Education Review B, 6*(11), 647–654.

Ayyagari, R. (2012). Impact of information overload and task-technology fit on technostress. Paper presented at the *Proceedings of the Southern Association for Information Systems Conference* (pp. 18–22).

Ayyagari, R., Grover, V., & Purvis, R. (2011). Technostress: Technological antecedents and implications. *MIS Quarterly, 35*, 831–858.

Bansal, S. (2020). Impact of the COVID-19 pandemic on education, rise of online teaching learning process & effects on health of kids. *Rise of Online Teaching Learning Process Effects on Health of Kids* (May 8, 2020).

Boelens, R., De Wever, B., & Voet, M. (2017). Four key challenges to the design of blended learning: A systematic literature review. *Educational Research Review, 22*, 1–18.

Bradley, G., & Russell, G. (1997). Computer experience, school support and computer anxieties. *Educational Psychology, 17*(3), 267–284.

Brush, T., Glazewski, K. D., & Hew, K. F. (2008). Development of an instrument to measure preservice teachers' technology skills, technology beliefs, and technology barriers. *Computers in the Schools, 25*(1–2), 112–125.

Chatterjee, S., & Bhattacharjee, K. K. (2020). Adoption of artificial intelligence in higher education: A quantitative analysis using structural equation modelling. *Education Information Technologies, 25*(5), 3443–3463.

Chatzoglou, P. D., Sarigiannidis, L., Vraimaki, E., & Diamantidis, A. (2009). Investigating Greek employees' intention to use web-based training. *Computers Education Information Technologies, 53*(3), 877–889.

Cheung, G. W., & Wang, C. (2017). Current approaches for assessing convergent and discriminant validity with SEM: Issues and solutions. Paper presented at the *Academy of Management Proceedings.*

Christian, M., Purwanto, E., & Wibowo, S. (2020). Technostress creators on teaching performance of private universities in Jakarta during Covid-19 pandemic. *Technology Reports of Kansai University, 62*(6), 2799–2809.

Covid, C. D. C., Team, R., Covid, C. D. C., Team, R., COVID, C., Team, R., ... & Skoff, T. (2020). Coronavirus disease 2019 in children—United States, February 12–April 2, 2020. *Morbidity and Mortality Weekly Report, 69*(14), 422.

Dong, Y., Xu, C., Chai, C. S., & Zhai, X. (2019). Exploring the structural relationship among teachers' technostress, technological pedagogical content knowledge (TPACK), computer self-efficacy and school support. *The Asia-Pacific Education Researcher, 1–11.*

Drent, M., & Meelissen, M. (2008). Which factors obstruct or stimulate teacher educators to use ICT innovatively? *Computers Education Information Technologies, 51*(1), 187–199.

Duboc, L., Kinnear, L., Alsina-Pagès, R. M., & Alías, F. (2019). Bringing Together Teaching, Research and Communication.https://www.researchgate.net/profile/Rosa-Alsina.Pages/publication/335608923_Bringing_Together_Teaching_Research_and_Communication_Initiatives_undertaken_by_the_GTM/links/5d6fcae992851cacdb214b6c/Bringing-Together-Teaching-Research-and-Communication-Initiatives-undertaken-by-the-GTM.pdf

Dulock, H. L. (1993). Research design: Descriptive research. *Journal of Pediatric Oncology Nursing, 10*(4), 154–157.

Eastin, M. S., & LaRose, R. (2000). Internet self-efficacy and the psychology of the digital divide. *Journal of Computer-Mediated Communication, 6*(1), JCMC611.

Effiyanti, T., & Sagala, G. H. (2018). Technostress among teachers: A confirmation of its stressors and antecedent. *International Journal of Education Economics and Development, 9*(2), 134–148.

Fahimirad, M., & Kotamjani, S. S. (2018). A review on application of artificial intelligence in teaching and learning in educational contexts. *International Journal of Learning Development, 8*(4), 106–118.

Ferguson, R., Clow, D., Beale, R., Cooper, A. J., Morris, N., Bayne, S., & Woodgate, A. (2015). Moving through MOOCS: Pedagogy, learning design and patterns of engagement. In: Conole G., Klobučar T., Rensing C., Konert J., & Lavoué E. (eds.) *Design for Teaching and Learning in a Networked World* (70–84): Cham: Springer.

Fernández-Batanero, J.-M., Román-Graván, P., Reyes-Rebollo, M.-M., & Montenegro-Rueda, M. (2021). Impact of educational technology on teacher stress and anxiety: A literature review. *International Journal of Environmental Research Public Health, 18*(2), 548.

Foulger, T. S., Graziano, K. J., Schmidt-Crawford, D., & Slykhuis, D. A. (2017). Teacher educator technology competencies. *Journal of Technology Teacher Education, 25*(4), 413–448.

Funminiyi, J., Akinlolu, A., & Agboola, P. (2014). Assessment of ergonomic hazards and techno-stress among the workers of Obafemi Awolowo University, Ile-Ife, Osun State, Nigeria. *Australian Journal of Business Management Research, 4*(1), 27–34.

Graham, C. R., Borup, J., & Smith, N. B. (2012). Using TPACK as a framework to understand teacher candidates' technology integration decisions. *Journal of computer Assisted learning, 28*(6), 530–546.

Grossman, J. H., Reid, P. P., & Morgan, R. P. (2001). Contributions of academic research to industrial performance in five industry sectors. *The Journal of Technology Transfer, 26*(1–2), 143–152.

Gupta, S. B., & Gupta, M. (2020). Technology and E-learning in higher education. *International Journal of Advanced Science and Technology, 29*(4), 1320–1325.

Hair, J. F., Money, A. H., Samouel, P., & Page, M. J. E. T. (2007). Research methods for business.

Hair Jr., J. F., Hult, G. T. M., Ringle, C., & Sarstedt, M. (2016). *A Primer on Partial Least Squares Structural Equation Modeling (PLS-SEM).* Thousand Oaks, CA, Sage Publications.

Hair Jr., J. F., Sarstedt, M., Hopkins, L., & Kuppelwieser, V. G. (2014). Partial least squares structural equation modeling (PLS-SEM).

Hair Jr., J. F., Sarstedt, M., Ringle, C. M., & Gudergan, S. P. (2017). *Advanced Issues in Partial Least Squares Structural Equation Modeling.* Thousand Oaks, CA: Sage Publications.

Hasan, S. M., Rehman, A., & Zhang, W. (2021). Who can work and study from home in Pakistan: Evidence from a 2018–19 nationwide household survey. *World Development, 138*, 105197.

Hatlevik, I. K., & Hatlevik, O. E. (2018). Examining the relationship between teachers' ICT self-efficacy for educational purposes, collegial collaboration, lack of facilitation and the use of ICT in teaching practice. *Frontiers in Psychology, 9*, 935.

Henseler, J., Ringle, C. M., & Sinkovics, R. R. (2009). The use of partial least squares path modeling in international marketing. In: Sinkovics R. R. & Ghaur P. N. (Eds.). *New Challenges to International Marketing.* Bingley: Emerald Group Publishing Limited.

Hox, J. J., & Boeije, H. R. (2005). Data collection, primary versus secondary.

Hsu, C.-L., & Lin, J. C.-C. (2018). Exploring factors affecting the adoption of internet of things services. *Journal of Computer Information Systems, 58*(1), 49–57.

Ibrahim, H., & Yusoff, Y. M. (2015). User characteristics as antecedents of techno stress towards EHRM: From experts' views. *Procedia-Social Behavioral Sciences, 172*, 134–141.

Jena, R. (2015). Technostress in ICT enabled collaborative learning environment: An empirical study among Indian academician. *Computers in Human Behavior, 51*, 1116–1123.

Joo, Y. J., Lim, K. Y., & Kim, N. H. (2016). The effects of secondary teachers' technostress on the intention to use technology in South Korea. *Computers Education Information Technologies, 95*, 114–122.

Kamal, N., Rabbani, S., Samdani, H., Shujaat, S., & Ahmad, M. (2020). Social media usage, overload and exhaustion: A performance perspective. *International Review of Management Marketing, 10*(5), 19–26.

Karr-Wisniewski, P., & Lu, Y. (2010). When more is too much: Operationalizing technology overload and exploring its impact on knowledge worker productivity. *Computers in Human Behavior, 26*(5), 1061–1072.

Kay, R. H. (2008). Exploring the relationship between emotions and the acquisition of computer knowledge. *Computers Education Information Technologies, 50*(4), 1269–1283.

Kena, G., Musu-Gillette, L., Robinson, J., Wang, X., Rathbun, A., Zhang, J., & Velez, E. (2014). US Department of Education, National Center for Education Statistics.

Kim, K. J. (2016). Interacting socially with the Internet of Things (IoT): Effects of source attribution and specialization in human–IoT interaction. *Journal of Computer-Mediated Communication, 21*(6), 420–435.

Kline, E., Wilson, C., Ereshefsky, S., Tsuji, T., Schiffman, J., Pitts, S., & Reeves, G. J. S. R. (2012). Convergent and discriminant validity of attenuated psychosis screening tools. *Schizophrenia Research, 134*(1), 49–53.

Kock, N. (2015). Common method bias in PLS-SEM: A full collinearity assessment approach. *International Journal of e-Collaboration, 11*(4), 1–10.

Kumar, S. (2019). Artificial intelligence divulges effective tactics of top management institutes of India. *Benchmarking: An International Journal, 26*(7), 2188–2204.

Lam, S.-F., Cheng, R. W.-Y., & Choy, H. C. (2010). School support and teacher motivation to implement project-based learning. *Learning instruction, 20*(6), 487–497.

Laurillard, D. (2007). Modelling benefits-oriented costs for technology enhanced learning. *Higher Education, 54*(1), 21–39.

Levie, J. (2014). The university is the classroom: Teaching and learning technology commercialization at a technological university. *The Journal of Technology Transfer, 39*(5), 793–808.

Li, L., & Wang, X. (2020). Technostress inhibitors and creators and their impacts on university teachers' work performance in higher education. *Cognition, Technology & Work, 23*, 1–16.

Liu, L., Chen, W., Solanas, A., & He, A. (2017). Knowledge, attitude, and practice about internet of things for healthcare. Paper presented at the *2017 International Smart Cities Conference (ISC2)*, IEEE, 1–4.

López-Vargas, O., Duarte-Suárez, L., & Ibáñez-Ibáñez, J. (2017). Teacher's computer self-efficacy and its relationship with cognitive style and TPACK. *Improving Schools, 20*(3), 264–277.

Majeed, A., & Ali, M. (2018). How Internet-of-Things (IoT) making the university campuses smart? QA higher education (QAHE) perspective. Paper presented at the *2018 IEEE 8th Annual Computing and Communication Workshop and Conference (CCWC)*, IEEE, 646–648.

Mehmood, T., Liland, K. H., Snipen, L., & Sæbø, S. (2012). A review of variable selection methods in partial least squares regression. *Chemometrics and Intelligent Laboratory Systems, 118*, 62–69.

Merriam, S. B., & Grenier, R. S. (2019). *Qualitative Research in Practice: Examples for Discussion and Analysis*. Hoboken, NJ: John Wiley & Sons.

Mirete, A. B., Maquilón, J. J., Mirete, L., & Rodríguez, R. A. J. S. (2020). Digital competence and university teachers' conceptions about teaching. *A Structural Causal Model, 12*(-12), 4842.

Mishra, P., & Koehler, M. J. (2006). Technological pedagogical content knowledge: A framework for teacher knowledge. *Teachers College Record, 108*(6), 1017–1054.

Mumtaz, N., Saqulain, G., & Mumtaz, N. (2021). Online academics in Pakistan: COVID-19 and beyond. *Pakistan Journal of Medical Sciences, 37*(1), 283.

Munro, B. H. (2005). *Statistical Methods for Health Care Research* (Vol. 1): Philadelphia, PA: Lippincott Williams & Wilkins.

Ocaña-Fernández, Y., Valenzuela-Fernández, L. A., & Garro-Aburto, L. L. (2019). Artificial Intelligence and Its Implications in Higher Education. *Journal of Educational Psychology-Propositos y Representaciones, 7*(2), 553–568.

Otieno, O. C., Liyala, S., Odongo, B. C., & Abeka, S. O. (2016). Theory of reasoned action as an underpinning to technological innovation adoption studies.

Patel, P., & Cassou, D. (2015). Enabling high-level application development for the Internet of Things. *Journal of Systems Software, 103*, 62–84.

Pavel, A.-P., Fruth, A., & Neacsu, M.-N. (2015). ICT and e-learning–catalysts for innovation and quality in higher education. *Procedia economics finance, 23*, 704–711.

Perez, J. A., Deligianni, F., Ravi, D., & Yang, G.-Z. (2018). Artificial intelligence and robotics. *arXiv preprint arXiv:10813.*

Phillips, J. (2012). Preparing teachers and developing school leaders for the 21st century. *OECD Publishing. Retrieved September, 9*, 2017.

Radha, R., Mahalakshmi, K., Kumar, V. S., & Saravanakumar, A. (2020). E-Learning during lockdown of Covid-19 pandemic: A global perspective. *International Journal of Control Automation, 13*(4), 1088–1099.

Rezaei, M., Mohammadi, H. M., Asadi, A., & Kalantary, K. (2008). Predicting e-learning application in agricultural higher education using technology acceptance model. *Turkish Online Journal of Distance Education, 9*(1), 85–95.

Ringle, C. M., Wende, S., & Will, A. (2005). SmartPLS 2.0 (beta). In: Hamburg.

Rodrigues, H., Almeida, F., Figueiredo, V., & Lopes, S. L. (2019). Tracking e-learning through published papers: A systematic review. *Computers & Education, 136*, 87–98.

Ruiz, A. B. M., Sánchez, F. A. G., & Pina, F. H. (2015). Cuestionario para el estudio de la actitud, el conocimiento y el uso de TIC (ACUTIC) en Educación Superior. Estudio de fiabilidad y validez. *Revista Interuniversitaria de Formación del Profesorado,* 83, 75–89.

Rushby, N., & Surry, D. (2016). *The Wiley Handbook of Learning Technology.* Hoboken, NJ: John Wiley & Sons.

Shehzadi, S., Nisar, Q. A., Hussain, M. S., Basheer, M. F., Hameed, W. U., & Chaudhry, N. I. (2020). The role of digital learning toward students' satisfaction and university brand image at educational institutes of Pakistan: A post-effect of COVID-19. *Asian Education Development Studies, 10*(2), 276–294.

Sánchez, J., Mallorquí, A., Briones, A., Zaballos, A., & Corral, G. (2020). An integral pedagogical strategy for teaching and learning IoT cybersecurity. *Sensors, 20*(14), 3970.

Semiz, K., & Ince, M. L. (2012). Pre-service physical education teachers' technological pedagogical content knowledge, technology integration self-efficacy and instructional technology outcome expectations. *Australasian Journal of Educational Technology, 28*(7).

Setiawan, A. R. (2020). Scientific literacy worksheets for distance learning in the topic of Coronavirus 2019 (COVID–19).

Shafique, K., Khawaja, B. A., Sabir, F., Qazi, S., & Mustaqim, M. (2020). Internet of things (IoT) for next-generation smart systems: A review of current challenges, future trends and prospects for emerging 5G-IoT scenarios. *IEEE Access, 8*, 23022–23040.

Shahzad, S. K., Hussain, J., Sadaf, N., Sarwat, S., Ghani, U., & Saleem, R. (2020). Impact of virtual teaching on ESL Learners' attitudes under COVID-19 circumstances at post graduate level in Pakistan. *English Language Teaching, 13*(9), 1–9.

Sheppard, B. H., Hartwick, J., & Warshaw, P. R. (1988). The theory of reasoned action: A meta-analysis of past research with recommendations for modifications and future research. *Journal of Consumer Research, 15*(3), 325–343.

Stentoft, D. (2017). From saying to doing interdisciplinary learning: Is problem-based learning the answer? *Active Learning in Higher Education, 18*(1), 51–61.

Tran, T., Ho, M.-T., Pham, T.-H., Nguyen, M.-H., Nguyen, K.-L. P., Vuong, T.-T., … Khuc, Q. (2020). How digital natives learn and thrive in the digital age: Evidence from an emerging economy. *Sustainability, 12*(9), 3819.

Tűrkeş, M. C., Căpuşneanu, S., Topor, D. I., Staraş, A. I., Hint, M. Ş., & Stoenica, L. F. (2020). Motivations for the use of IoT solutions by company managers in the digital age: A Romanian case. *Applied Sciences, 10*(19), 6905.

Wang, K., Shu, Q., & Tu, Q. (2008). Technostress under different organizational environments: An empirical investigation. *Computers in Human Behavior, 24*(6), 3002–3013.

Warshaw, P. R., & Davis, F. D. (1985). Disentangling behavioral intention and behavioral expectation. *Journal of experimental social psychology, 21*(3), 213–228.s

Weil, M. M., & Rosen, L. D. (1997). *Technostress: Coping with Technology@ Work@ Home@ Play*: New York: John Wiley & Sons, Vol. 13, p. 240.

Werang, B. R. (2018). The effect of workload, individual characteristics, and school climate on teachers' emotional exhaustion in elementary schools of Papua. *Cakrawala Pendidikan, 37*(3), 457–469.

Whyte, S. (2014). Bridging the gaps: Using social media to develop techno-pedagogical competences in pre-service language teacher education. *Recherche et pratiques pédagogiques en langues de spécialité. Cahiers de l'Apliut, 33*(2), 143–169.

Yin, C., Lim, K. H., Sun, Y., & Zhao, D. (2012). A technology-individual contingency perspective of mobile technostress: The moderating role of personality. Paper presented at the *18th Americas Conference on Information Systems 2012*, AMCIS 2012.

7 Effects of Digital Game-Based Learning and Traditional Teaching Approaches on Students' Mathematics Problem-Solving Attitude

Nadira Dayo
The Begum Nusrat Bhutto Women University Sukkur

Muhammad Mujtaba Asad and Unaeza Alvi
Sukkur IBA University

CONTENTS

DOI: 10.1201/9781003143796-7

7.1 INTRODUCTION

Problem-solving is an essential human cognitive process that helps to succeed in the 21st century. Most problem-solving skills are learned in mathematics-related subjects. Despite this, most students avoid this subject and consider it a dull and boring subject. Consequently, their attitude toward mathematics, particularly for problem-solving, becomes negative. Conversely, utilizing technology, mainly digital mathematics games, can help develop a favorable attitude for mathematics problem-solving. Attitude is the main element in the academic success of students (Anghelache, 2013); therefore, students' attitude should be studied to alter it at the beginning stage before they give up studying a vital subject, mathematics, and avoid learning the necessary skill, problem-solving. If their attitudes are not altered, then they are less likely to continue their education and perform assigned educational tasks.

The educational environment influences the attitude of students. In traditional classrooms, the brain's activity is almost zero, and students only grow in size, but they do not acquire other educational skills such as reasoning, critical thinking, problem-solving, creativity, and confidence. New teaching and learning trends have emerged in the last few decades, and gamified learning is one such. Compared with conventional teaching, the digital game-based learning (DGBL) teaching approach was found to be better and effective (Boeker, Andel, Vach, & Frankenschmidt, 2013). DGBL increases students' mathematical confidence and performance (Foster, 2016), conceptual knowledge (Denham, 2019), retention (Vogt, 2018), and overall attitude towards mathematics problem-solving (Nizaruddin, Muhtarom, & Sugiyanti, 2017; Pratama & Setyaningrum, 2018). Regardless of knowing the importance of DGBL in enhancing students' attitudes towards mathematics problem-solving, limited studies have been conducted.

Responding to mixed opinions about mathematics problem-solving and lack of evidence about the influence of the DGBL teaching approach on mathematics problem-solving attitude, in the current study, attitudes of K-8 grade students in traditional and DGBL classrooms were compared. Thus, the current study's main aim was to compare the effects of traditional and DGBL classrooms on the mathematics problem-solving attitude of K-8 grade students.

7.2 RESEARCH QUESTIONS

RQ1: What is the difference in K-8 grade students' mathematics problem-solving attitude in traditional and digital game-based learning classrooms?

RQ2: Is there any statistically significant difference in the mathematics problem-solving attitude of the experimental (taught in DGBL classroom) and control group (traditionally taught) students?

7.3 LITERATURE REVIEW

7.3.1 Difference Between Traditional and Digital Game-Based Teaching Approach

Numerous research has identified the advantages of DGBL over the conventional teaching approach (Liao, 2010; Vogt, 2018). In traditional classrooms, students' creative thinking is not encouraged, while in DGBL classrooms, students think differently (Behnamnia, Kamsin, Ismail, & Hayati, 2020). Likewise, the theory is taught to learners in a traditional class by just telling them, but in DGBL, students apply the learning in a virtual environment and becomc able to use those skills in real life (Lesser, 2020). Furthermore, there is a difference in cost for traditional and DGBL teaching. For conventional education, the school needs to have books, stationery, and papers to teach and assess students annually. On the other hand, for DGBL, once the infrastructure is established, it can be used for many years. Most importantly, learning through DGBL is fun for students, and they get engaged in learning without being frustrated. In contrast, this element is missing in traditional classrooms as students are expected to listen and reproduce content. Similarly, digital games can be accessed anywhere, and once students' attitude toward mathematics learning is developed, they can learn through digital games anywhere, unlike traditional classrooms where a teacher is needed to motivate students for learning (Papadakis, 2018)

7.3.2 Effects of DGBL Teaching Approach on Attitude towards Mathematics Problem-Solving

Digital games are no longer viewed as merely entertaining tools, but they have been regarded as an educational tool that can positively affect students' mathematics problem-solving attitudes. Furthermore, challenging subjects such as mathematics can be made enjoyable by teaching through digital games as, in the games, interactive and responsive instructions are provided. Moreover, researchers have claimed that digital games are appropriate for today's learners, whose learning attitudes and interests, and learning motivation are affected through computerized games and technologies (Noraddin & Kian, 2014). The learners of the 21st century cannot passively absorb the knowledge given by teachers because of advancements in technology since they get exposure to technical devices, have access to knowledge at their fingertips, and use social media from an early age. According to Dayo, Alvi, and Asad (2020), a mathematics problem-solving attitude comprises confidence to solve mathematics problems, remain persistent while solving problems, have an interest, establish a connection of mathematics problems with real-life, and make sense of mathematics problems. Each of these components of mathematics problem-solving attitude and their association with DGBL is discussed in the coming sections.

7.3.3 DGBL Improves Mathematics Problem-Solving Confidence

The quality of learning can be improved by making students confident. Earlier studies have demonstrated a strong relation of confidence with the performance of students in the learning of mathematics as well (Foster, 2016). Not only this, but studies have found a close link between self-confidence and choice for enrolling in the mathematics field (Ballane, 2019); it has also been noticed that due to lack of confidence in mathematics, very few students pursue Science, Technology, Engineering and Mathematics (STEM) education fields. Furthermore, confidence helps students decide how much effort they will put into learning; any student who lacks confidence will not put much efforts to learn. Given this, digital game-based teaching as a student-centered pedagogy fosters the confidence and autonomy of students. In addition, teaching through games reduces the mathematics anxiety of students and provides an opportunity for them to practice mathematics problems at their own pace. Another advantage of using digital mathematics games is that students receive immediate feedback and instant reward in digital games. Gradually, their confidence improves, and rational thinking is developed (Ahmad, Malik, Siddiqui, & Khan, 2018).

7.3.4 DGBL Encourages Mathematics Problem-Solving Persistence

Persistence is a crucial skill for mathematics problem-solving, and DGBL supports the learning of this skill (Van Eck, 2015). In a study conducted by Ventura, Shute, and Zhao (2013), it was shown that learners involved in playing digital games exhibited more perseverance in solving problems than those who were not given a chance to play digital games. Correspondingly, Case (2013) discussed that digital games helped students improve their retention and persistence; therefore, they could be used to make students tackle different mathematics problems and solve real-life problems. In the long run, persistence reduces dropout rates of students. By teaching through digital games, students learn interactively; hence, they become more responsive and show persistence to solve even hard and difficult problems since problems are presented in the game form (Israel-Fishelson & Hershkovitz, 2020).

7.3.4.1 Students' Interest Develop by Learning through DGBL

The majority of mathematics teachers find it challenging to develop students' interest in mathematics problem-solving. Research suggests that DGBL increases students' interest in learning mathematics problem-solving (Dayo et al., 2020) by diversifying education experiences for students. In another study by Yeh, Cheng, Chen, Liao, and Chan (2019), it was found that students' mathematics problem-solving achievement was improved by playing digital games, and there is a strong correlation between achievement and interest in mathematics. By engaging students in different learning activities through games, desired learning outcomes can be achieved, and students' interest also increases in mathematics learning. As digital game-based teaching is a novel and innovative pedagogy, it triggers students curiosity and interest without making them realize that they are being taught.

7.3.4.2 DGBL Support Real-World Application of Mathematics Problem-Solving

DGBL is a constructive teaching approach that enables students to learn in the virtual world and transfer that into real life; hence, they can avoid mistakes that can possibly take place in an actual context. Additionally, by teaching through digital mathematics games, students grasp abstract concepts and establish connections between the real and technological world. As a result, students turn out to be more practical and effective problem-solvers. In the DGBL teaching approach, students are exposed to the virtual environment, which is also a simulation of the real world; therefore, after being taught through digital games, students learn to represent and understand mathematical concepts in real life (Lowrie & Jorgensen, 2015, p. 144). Chang, Wu, Weng, and Sung (2012), by emphasizing the same thing, stated that when students learn through digital games, they imitate the virtual environment teaching in real life, which results in a collaborative and authentic learning experience.

7.3.4.3 Students Learn to Make Sense of Mathematics Problems by DGBL

Higher-order cognitive abilities and skills are required to make sense of mathematics problems. One of the promises of DGBL is that it equips students with 21st-century skills, and making sense of mathematics problems can be considered one of them (Gee, 2003; Squire, 2008). Lowrie and Jorgensen (2015) discuss that since digital games help students visualize problems; therefore, they become able to make sense of mathematics problems in a better way.

The above review of the literature suggests that DGBL can improve students' attitudes in mathematics problem-solving. However, the empirical instances are scarce, mainly in Pakistan's context; therefore, it is reasonable to research the digital game-based teaching approach's influence on mathematics problem-solving attitude and compare it with the traditional teaching approach.

7.4 RESEARCH METHODOLOGY

7.4.1 Research Design

The study's primary purpose was to compare the effects of traditional and DGBL teaching approaches on the mathematics problem-solving attitude of K-8 grade students. Therefore, to carry out the study, a true experimental research design was followed.

7.4.2 Research Participants

Since it was a true experimental research design, randomly, participants were allocated into the controlled and experimental groups. There were 32 K-8 grade students in each group. The experimental group comprised 20 males and 12 females, whereas the control group included 9 females and 23 males.

7.4.3 Data Collection Tool

The quantitative tools and pre–post surveys were utilized to collect data about the effects which traditional and digital game-based teaching approaches have on the mathematics problem-solving attitude of K-8 grade students for the current research. The researcher developed the survey since a survey on students' mathematics problem-solving attitude was not available.

TABLE 7.1
Reliability Statistics and Distribution of Students' Mathematics Problem-Solving Attitude Survey

Students' Mathematics Problem-Solving Attitude Survey	No. of Items	Cronbach's Alpha Value
Confidence	4	0.823
Sense-making	6	0.774
Persistence	4	0.774
Interest	3	0.763
Real world connection	3	0.814
Overall survey items	20	0.809

The students' attitude survey underwent several subsequent revisions before it was administered. It was one of the difficult things to determine what constitutes a mathematics problem-solving attitude. However, after a rigorous study of literature, few recurring aspects of students' attitudes were selected, namely confidence, sense-making, real-world connection, interest, and persistence. A five-point Likert scale was used to record students' responses, with one indicating strongly disagree, and five, strongly agree. The total number of items of each aspect of students' mathematics problem-solving attitude and Cronbach's alpha value, which was calculated to identify whether the tool was reliable, is shown in Table 7.1.

From the above table, it can be seen that Cronbach's alpha value of the overall survey was .809, which is considered an excellent value and establishes the reliability of the survey.

7.4.4 Data Collection Procedure

A 1-month experiment was done in which students of both groups were taught for two periods of 90 minutes daily. The experimental group was educated by using the DGBL teaching approach. In contrast, the traditional teaching approach was used to teach the control group. The existing digital mathematics games were used to carry out the experiment. However, they were aligned with the school syllabus. Furthermore, before teaching both groups, a presurvey (students' attitude towards mathematics problem-solving) was conducted. Following the presurvey, randomly, participants were allocated to controlled and experimental groups.

The experimental group students were taught in a computer lab, and they were organized in groups or with peers. In 45 minutes, the first 10 minutes were spent on eliciting prior knowledge, communicating learning objectives, and stating gameplay rules to students. After that, students were allowed to play digital mathematics games for 20–25 minutes. During gameplay, the teacher used to assist and guide students in case they feel any confusion. The feedback was provided to students in games, and they were assessed by doing postgame debriefing, keeping in-game feedback evidence, and taking tests after the gameplay. Lastly, learning was summarized by

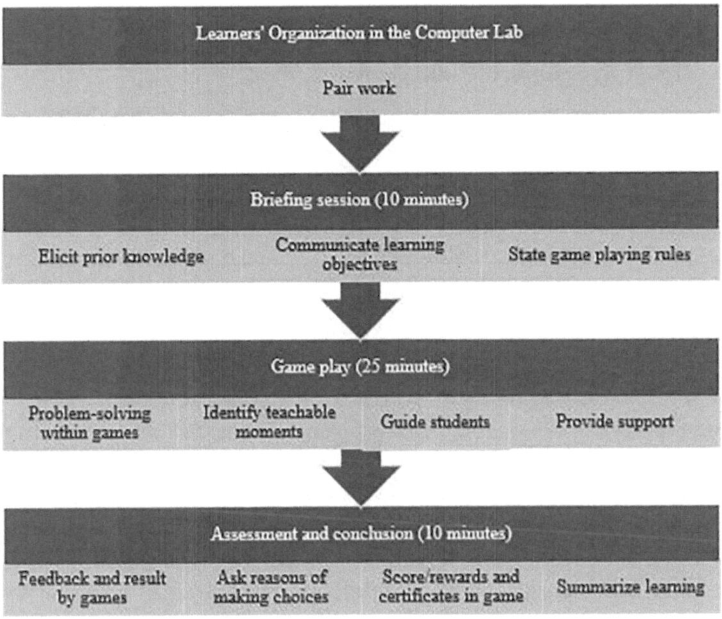

FIGURE 7.1 Digital games for problem-solving attitude.

asking students what they learned. Figure 7.1 demonstrates the adopted way of teaching through digital games for improving mathematics problem-solving attitude.

In contrast, the control group was taught in the classroom, and they were organized to complete assigned tasks alone. The teacher used to teach them by writing on board and asking questions from students rarely. The students used to write solutions to problems in their copy book. Unlike the DGBL classroom, the assessment of students was done by taking the paper-pencil test only. Two units (operations on sets and quadratic equations) were taught to both groups of students from their prescribed syllabus. Finally, a postsurvey of mathematics problem-solving attitude was conducted.

7.4.5 Data Analysis

The data were analyzed using Statistical Package for the Social Sciences (SPSS) 24. Descriptive statistics and Cronbach's alpha value was calculated. Furthermore, the data normality was also checked by generating a histogram and doing the Shapiro-Wilk test. The normality of data was determined by the results of Shapiro-Wilk's tests of normality.

Table 7.2 shows that the p-value of the experimental group in the mathematics problem-solving attitude survey was $p = .075$. The p-value of the control group mathematics problem-solving attitude survey was $p = 0.899$. Significance value $p > .05$ showed that the data were normal. Moreover, to know whether the difference in experimental and control group students' mathematics problem-solving attitude is significant, an independent-sample t-test was done. Finally, the eta square was calculated to determine the effect size.

TABLE 7.2
Results of the Shapiro-Wilk Normality Test

	Group	Shapiro-Wilk		
		Statistic	df	Sig.
Students' Mathematics Problem-Solving Attitude Survey	Experimental	.906	18	.075
	Control	.969	12	.899

7.5 RESULTS

In Table 7.3, prepost survey results of both groups' students, experimental and control groups, are reported in the mathematics problem-solving attitude survey. In the first studied aspect of mathematics attitude, i.e., confidence, the presurvey mean score of the experimental group was 1.90 and control was 2.77. In postsurvey, it increased for both groups, experimental, to 3.51 and control, to 3.10. The mean score of the experimental group in confidence has improved more than that of the controlled group.

In the sense-making aspect, the experimental group's mean score in presurvey was 2.90, and the control group 2.63. Like previous results, in postsurvey, the mean score of both groups increased. Besides, when compared with control group, again, the experimental group's means score increased significantly.

Furthermore, Table 7.3 shows that the experimental group's persistence aspect improved more than the control group. The experimental group's mean score in presurvey was 2.07. The mean score of control group in presurvey was 2.21; however, the postsurvey results were as follows: experimental 2.92 and control group 2.57.

In the interest aspect of mathematics problem-solving attitude, the students' mean scores were not significantly different. The pre–post survey results of both groups were: experimental group, pre=2.61, post=3.10, and control group, pre=2.68, and post=3.00. These results show that the interest level of both groups was nearly the same. Nevertheless, again experimental group's interest, according to survey results, was slightly better than that of the control group.

Finally, the surprising results can be seen in the real-world connection aspect of mathematics problem-solving attitude as the experimental group's mean scores are significantly higher than all other aspects of mathematics problem-solving attitude. The experimental group presurvey score was 2.32, and the postsurvey was 3.38. The control group score in presurvey was 2.48, and postsurvey, it was 2.64. From this, it can be concluded that digital mathematics games enable students to apply the learning and establish the connection of mathematics with real life.

Looking at all mean score differences in presurvey to postsurvey of experimental and control groups, it is concluded that students who learn mathematics in the DGBL classroom have a positive attitude towards mathematics problem-solving as their confidence, sense-making, persistence, interest, and real-world connection with mathematics problem-solving was higher than those students who learned in the traditional mathematics classroom. Moreover, whether this difference is significant or not is explained through the results of the independent-sample t-test in Table 7.4.

TABLE 7.3
Descriptive Statistics of Students' Mathematics Problem-Solving Attitude Survey

Students' Mathematics Problem-Solving Attitude Survey	Time	Group	Mean	SD
Confidence	Presurvey	Experimental	1.90	1.06
		Control	2.77	.76
	Postsurvey	Experimental	3.51	.66
		Control	3.10	1.03
Sense-making	Presurvey	Experimental	2.90	.99
		Control	2.63	.78
	Postsurvey	Experimental	3.20	.81
		Control	2.95	.79
Persistence	Presurvey	Experimental	2.07	1.00
		Control	2.21	1.02
	Postsurvey	Experimental	2.92	.99
		Control	2.57	1.03
Interest	Presurvey	Experimental	2.61	1.19
		Control	2.68	1.25
	Postsurvey	Experimental	3.10	1.17
		Control	3.00	1.03
Real-world connection	Presurvey	Experimental	2.32	1.06
		Control	2.48	1.15
	Postsurvey	Experimental	3.38	.94
		Control	2.64	1.25

$N=64$ (32 experimental and 32 control).

TABLE 7.4
Results of Independent-Sample T-Test of Students' Mathematics Problem-Solving Attitude Survey

Students' Mathematics Problem-Solving Attitude Survey	Group	Mean	SD	t	Sig (2-tailed)	Effect Size
Presurvey	Experimental	2.36	0.56	1.47	0.14	0.01
	Control	2.56	0.50			
Postsurvey	Experimental	3.22	0.63	2.47	0.01	0.09
	Control	2.85	0.56			

$N=64$ (32 experimental and 32 control).

It can be seen in Table 7.4 that in the presurvey, the experimental group's mean score was 2.36, and it was 2.56 for the control group. Whereas postsurvey results revealed that the experimental group's mean score significantly increased to 3.22, and the control group's mean score also slightly increased to 2.85. Furthermore, postsurvey results show a statistically significant difference in experimental and control group mathematics problem-solving attitude, $t(64)=2.47$, $p=0.01$. Moreover, the effect size in the presurvey mean scores was very small (eta squared=0.01). In the postsurvey, the magnitude of the experiment effect was large (eta squared=0.09), which indicates that digital mathematics games significantly improve students' attitudes towards learning mathematics problem-solving.

7.6 DISCUSSION

The primary purpose of the current research was to compare the difference in mathematics problem-solving attitudes of K-8 grade students who learned through two different teaching approaches that are DGBL and traditional. It was an experimental study in which the data were collected by taking pre- and postsurvey of mathematics problem-solving attitude. The survey contained the following components of mathematics problem-solving attitude: confidence in solving mathematics problems, making sense of mathematics problems, persistence in solving mathematics problems, interest in solving mathematics problems, and establishing a connection of mathematics problems with real life.

The experimental group, taught by the DGBL teaching approach, showed a significant increase in mean score in the mathematics problem-solving attitude survey (presurvey 2.36 and postsurvey 3.22). In contrast, the control group (taught by traditional teaching approach) had mean scores that were lower (presurvey 2.56 and postsurvey 3.10). The difference between the experimental and control groups' mean scores was apparent, and it was also statistically significant. It can be concluded that DGBL is an active learning approach that improves students' mathematics problem-solving attitude. This is because digital games make students confident, help them make sense of mathematics problems, help them persistently solve mathematics problems, develop interest, and connect mathematics problems with real life.

In the current research, the experimental group students were permitted to work in groups and solve problems using the trial-and-error approach, which digital games offer. Furthermore, students visualized mathematics problems through games, which resulted in their interest developing for learning mathematics problem-solving. Likewise, they persisted in solving digital game problems and won the reward (Yeh et al., 2019). These findings are parallel with the results of other studies (Çankaya & Karamete, 2009; Starkey, 2013). In other studies (Çankaya & Karamete, 2009; Mavridis, Katmada, & Tsiatsos, 2017), a positive impact of digital games on experimental group students' attitude towards mathematics was found in contrast to the control group students. All in all, it was revealed that the students who were exposed to digital mathematics games had a favorable attitude for mathematics problem-solving than their counterparts who were not exposed to digital mathematics games.

7.7 CONCLUSION

The research has contributed to the literature by providing evidence of the effects of DGBL on students' mathematics problem-solving attitude. An important conclusion drawn from the current study was that the DGBL teaching approach significantly improves students' attitude toward mathematics problem-solving compared to traditional teaching approaches. Nonetheless, the present study has certain limitations. First, merely determining the effects of DGBL on attitude is not enough; therefore, mathematics problem-solving achievement and skills component should have also been studied. Second, the other grade students, such as primary, secondary, and higher secondary students, could have also been included, and a comparison could have been made. Due to limited time and resources, it was not done in current study. Finally, the prolonged effects of DGBL could have been determined with a longitudinal research design, which was not done. In light of these limitations, the current research has provided a base to conduct more research on the effects of the DGBL teaching approach on mathematics problem-solving. Based on findings, it is recommended that the teachers should utilize a digital game-based teaching approach in mathematics classrooms since the effects are positive. Parents should invest in buying effective digital games to improve the learning of mathematics problem-solving. Finally, teacher education programs should include this innovative pedagogy, DGBL, to equip students with problem-solving skills in their teacher education program.

REFERENCES

Ahmad, F. H., Malik, M., Siddiqui, S., & Khan, H. (2018) Investigating the Impact of Game-Based Learning in Mathematics Using Tablets Among Primary School Students.

Anghelache, V. (2013). Determinant factors of students' attitudes toward learning. *Procedia-Social and Behavioral Sciences, 93*, 478–482.

Ballane, G. (2019). Understanding of Self-Confidence in High School Students.

Behnamnia, N., Kamsin, A., Ismail, M. A. B., & Hayati, A. (2020). The effective components of creativity in digital game-based learning among young children: A case study. *Children and Youth Services Review, 116*, 105227.

Boeker, M., Andel, P., Vach, W., & Frankenschmidt, A. (2013). Game-based e-learning is more effective than a conventional instructional method: a randomized controlled trial with third-year medical students. *PloS one, 8*(12), e82328.

Çankaya, S., & Karamete, A. (2009). The effects of educational computer games on students' attitudes towards mathematics course and educational computer games. *Procedia-Social and Behavioral Sciences, 1*(1), 145–149.

Case, R. E. (2013). The Impact of Digital Games on Student Persistence and Retention in an Online Higher Education Context. *ProQuest LLC.*

Chang, K.-E., Wu, L.-J., Weng, S.-E., & Sung, Y.-T. (2012). Embedding game-based problem-solving phase into problem-posing system for mathematics learning. *Computers & Education, 58*(2), 775–786.

Dayo, N. A., Alvi, U., & Asad, M. M. (2020). *Mechanics of Digital Mathematics Games for Learning of Problem-Solving: An Extensive Literature Review.* Paper presented at the 2020 International Conference on Emerging Trends in Smart Technologies (ICETST).

Denham, A. R. (2019). Using the PCaRD digital game-based learning model of instruction in the middle school mathematics classroom: A case study. *British Journal of Educational Technology, 50*(1), 415–427.

Foster, C. (2016). Confidence and competence with mathematical procedures. *Educational Studies in Mathematics, 91*(2), 271–288.

Gee, J. P. (2003). What video games have to teach us about learning and literacy. *Computers in Entertainment (CIE), 1*(1), 20–20.

Israel-Fishelson, R., & Hershkovitz, A. (2020). Persistence in a game-based learning environment: The case of elementary school students learning computational thinking. *Journal of Educational Computing Research, 58*(5), 891–918.

Lesser, A. J. (2020). An investigation of digital game-based learning software in the elementary general music classroom. *Journal of Sound and Music in Games, 1*(2), 1–24.

Liao, Y.-K. (2010). *Game-based learning vs. traditional instruction: A meta-analysis of thirty-eight studies from Taiwan.* Paper presented at the Society for Information Technology & Teacher Education International Conference.

Lowrie, T., & Jorgensen, R. (2015). *Digital Games and Mathematics Learning: Potential, Promises and Pitfalls* (Vol. 4). Berlin: Springer.

Mavridis, A., Katmada, A., & Tsiatsos, T. (2017). Impact of online flexible games on students' attitude towards mathematics. *Educational Technology Research and Development, 65*(6), 1451–1470.

Nizaruddin, N., Muhtarom, M., & Sugiyanti, S. (2017). Improving students' problem-solving ability in mathematics through game-based learning activities.

Noraddin, E. M., & Kian, N. T. (2014). Academics' attitudes toward using digital games for learning & teaching in Malaysia. *Malaysian Online Journal of Educational Technology, 2*(4), 1–21.

Papadakis, S. (2018). The use of computer games in classroom environment. *International Journal of Teaching and Case Studies, 9*(1), 1–25.

Pratama, L., & Setyaningrum, W. (2018). *Game-Based Learning: The effects on student cognitive and affective aspects.* Paper presented at the Journal of Physics: Conference Series.

Squire, K. (2008). Video-game literacy: A literacy of expertise. In: Coiro, J., Knobel, M., Lankshear, C. & Leu, D. (Eds.) *Handbook of Research on New Literacies* (pp. 635–670): New York: Lawrence Erlbaum Associates.

Starkey, P. L. (2013). The effects of digital games on middle school students' mathematical achievement.

Van Eck, R. (2015). Digital game-based learning: Still restless, after all these years. *EDUCAUSE Review, 50*(6), 13.

Ventura, M., Shute, V., & Zhao, W. (2013). The relationship between video game use and a performance-based measure of persistence. *Computers & Education, 60*(1), 52–58.

Vogt, S. (2018). Middle School Teachers' Use and Perceptions of Digital Game-Based Learning.

Yeh, C. Y., Cheng, H. N., Chen, Z.-H., Liao, C. C., & Chan, T.-W. (2019). Enhancing achievement and interest in mathematics learning through Math-Island. *Research and Practice in Technology Enhanced Learning, 14*(1), 5.

8 Challenges and Futuristic Approach of Blended Learning in Higher Education

Prashant Gupta, Trishul Kulkarni, and Bhagwan Toksha
Maharashtra Institute of Technology

CONTENTS

8.1 INTRODUCTION

The change in the lifestyles, embracing newer and advanced technologies in the field of Information and Communication Technology (ICT), has led to major shifts in the higher education for the 21st century. One such major shift is the acceptance of a blended mode of teaching-learning across higher education with more flexible mixing of online and traditional classroom instruction, making the teaching-learning process easier for instructors and learners to communicate in non-traditional methods (Caner, 2012). The philosophical standpoint behind the same is to make the expertise of leading instructors available to a wider spectrum of students, overcoming

DOI: 10.1201/9781003143796-8

time-strain for the students to have access to the learning content, cater to the needs of tech-savvy students, and incorporate various instructional strategies. The evolution in online educational content creation and management has made blended learning a naturally developing phenomenon as per the educational purposes and contexts.

The term Blended Learning (BL) is placed arbitrarily between the extreme ends of traditional classroom teaching and online instruction, which resulted in the ambiguity of what, how, and when educators mix different delivery methods of instruction (Brown, 2015). The scholars in the field have embraced BL models as the new traditional model or the new normal model (Norberg et al., 2011; Ross & Gage, 2006). However, as the definition of BL has ambiguities in both design and implementation, tracking of the accurate extent of its growth is challenging (Oliver &Trigwell, 2005). Though there are many attempts to define BL, the effective inclusion of online tools for teaching-learning is what matters the most. The exact definition of BL, in its ultimate form, is a combination of online and face-to-face methodology (Watson, 2008). The different definitions of the term perceived by educationists along with the environment or context of application are summarized in Table 8.1.

The unforeseen situation of COVID-19 in the year 2020 has challenged the belief that the contact, accessibility, and interactions required by all human-nature processes including trade, teaching-learning, social gatherings, etc. will be available as an unlimited resource. As a result, it forced the stakeholders involved to use ICT as the main channel of interaction. More specifically, the face-to-face teaching-learning could not remain unaffected in such an unprecedented situation and the inclusion of online components in education got reasserted in a major way. Thankfully, the search for growth and finding alternatives for better teaching-learning and training practices have remained one of the major concerns of educationalists for a long time. The time-stamps of such experiments date back to 1840s when Sir Issac Pitman launched the first distance education course (Archibald & Worsley, 2019). The shorthand texts were sent to students via mailed postcards and they were instructed to send the postcards back to be graded and corrected. Computers and mobile devices were not available in those days and still, effective feedback and assessments could be achieved. The next major milestone was in the 1970s when the mainframe computer-based training was started. The training was deployed to workers within an organisation without having to rely on printed teaching-learning materials and face-to-face instruction. The regime of television in the 1980s enabled the use of video networks to train learners and learners used to watch the videos of instructors on television and communicate with their peers. The era of CD-ROM training and Learning Management Systems (LMS) brought another dimension to the blended mode of instructions in the 1990s. The institutions started using CD-ROMs for interactive learning experiences, and the first LMS with limited functionality (as compared to what today's LMS offers) was also introduced during these times. More recently, the 20th-century era could be recognized as the real modern version of the blended mode of teaching-learning in which the advent of the internet, ease in access of educational content, web-based assessments and assignments replaced the system of distributing CD-ROMs among learners. Hence, the blend of face-to-face instruction and technology-based learning is creating newer and more creative ways to enrich the educational experience, making the learning fun, exciting, and even more beneficial.

TABLE 8.1
Definitions of BL Along with the Environment or Context of Application

Sr. No.	Definition Perceived	Environment or Context of Application	Reference
1	Combination of face-to-face and online instruction	Higher education, k-12, and corporate training	Graham (2013)
2	Blend of an online component in terms of tutorials, CAD, and the traditional component in terms of laboratory and in-class activities using modelling compounds supplemented with workshop activities and field trips	Mechanical and Mechatronics Engineering students learning Arduino microcontroller	Jones and Chew (2015)
3	Optimized achievement of learning objectives with proper learning technologies considering personalized learning style to transfer the right skill at the right time	Employee orientation program	Singh and Reed (2001)
4	Mixing or combining instructional technology with intended tasks for creating a harmonious effect	Overall teaching-learning	Driscoll (2002)
5	An e-learning segment prior to a face-to-face conference, discussion, and debate at the conference	Overall teaching-learning	Ward and LaBranche (2003)
6	Significant content delivered online with a reduced number of face-to-face meetings	US higher education	Allen and Seaman (2010)
7	A combination of two or more of all possible learning types with at least one of the learning types must be a physical class-based type and at least one other learning type must be e-learning type	Overall teaching-learning	Kim (2007)
8	Hybrid model wherein the online component is provided to learners as a natural extension of traditional classroom learning, instruction occurring in the traditional classroom as well as online	Overall teaching-learning	Collis and Moonen (2001)

8.2 INSTRUCTOR–STUDENT INTERACTIONS

The BL methodology benefits the teaching-learning process by helping to access the knowledge gained with efficiently teaching the concepts of the said process. It improves the accessibility of education as the course content is not just limited to the classroom, which is the case with traditional teaching-learning methods. The students can learn at their own pace with games, applications, or programs as the learning becomes individualized and engaging instead of the "one instruction for all" concept, thereby improving satisfaction and reducing the cognitive load (tension) among the students. Islam et al. (2018) divided XI grade students into two learning groups with control group learning via the use of direct model and the experimental group via the blended model. It was found that the experimental group reported a significant improvement in their motivation and achievement with the use of a BL model. Kintu et al. (2017a) used multiple regression analysis to exhibit the interdependence of design features of BL such as technology, online tools, student attitude, background, etc. to predict the outcome as student satisfaction. Similar results were reported by Ginaya et al. (2018) for improvement in English speaking abilities along with motivation. Herbert et al. (2017) demonstrated a model for the application of BL in large group teaching sessions, for over 250 learners in a single learning group with the implementation of short online modules uploaded on "Moodle". Almost 50% of the face-to-face lectures were reduced to online modules and interactions with large-group sessions were conducted over "Echo360" active learning platform which recorded around 88% respondents. The modules promoted understanding along with the active learning of students at their own pace in an enjoyable and motivating manner. It can be understood as a fun concept as it relieves the students of the long boring lectures. It also increases their chances of pursuing higher education as it provides a positive experience for them in the former schooling (K12) years. The student-instructor interaction is the most important feature as the relationship is bound to benefit both the involved stakeholders to maintain progress and gain deeper knowledge by asking questions for instructor and student, respectively.

Interaction can be broadly defined as "a communication where multiple persons are involved." "Interaction" was first studied by Parker (1999) as an activity in BL and he defined it as a reciprocal communication over a specific context among the ones who send and receive. Berge (1999) defined the notion of interaction in BL as two-way communication between two or higher individuals aimed for problem-solving along with a societal relevance towards relationship building. In general, the interaction may be done for information sharing, problem-solving, or social exchange. However, there are reports (2000) that the scope of interaction may be tapered down to apply it when communications between the individuals are pointed towards problem-solving and/or exchange of information.

Bernard et al. (2014) drew a comparison of various forms of interactions that may occur to understand the best combination which can lead to the highest achievements in the BL model. He reported the focus on three major categories, i.e., instructor-student interactions, student-student interactions, and student-content interactions. In general, instructor-student interactions are blended as they happen over the conventional face-to-face methodology and the more advanced online mode such as video conference, chats, and emails. The student-student interaction happens between different learners within a small workgroup through joint assignments, projects, teams, etc. The student-content interaction is slightly different and happens while the student goes through the course content during both online and face-to-face teaching-learning modes. The aim is to construct meaning from the content, absorbing the knowledge, and applying it for problem-solving and related activities. It was learned that the consequences of student-student and student-content interaction were significantly better than student-instructor interaction and a surprising relationship between student-student with student-content interactions and student-instructor with student-content interactions were revealed. However, the results for a combination of student-student and student-instructor interaction were not the same as the above two. Also, it was found that the high- and medium-scale interactions helped higher achievements than low-scale interactions for all three classifications.

BL impacts the teaching and learning environment as it interacts with cognitive, affective, and behavioural components of students (Dziuban et al., 2018). In a BL environment, the instructor's cognitive role has a deeper complexity as new tools are required to express emotions regarding the affective role. The managerial role of the instructor necessitates detailing, structuring, and better student monitoring. The importance of a set of conditions to understand the instructor behavior and their eagerness/hesitancy to adopt information and communication technology was reported by Demetriadis et al. (2003). The instructors think that ICT augments their image and the teaching quality, mitigates disturbances that may disrupt the action leading to issues for achieving curriculum goals, and instils confidence in the use of various tools or dissemination of course content which may raise issues in instructor-student and instructor-institution relationship. In case of a past experience with technology's use in the teaching-learning process and the knowledge accumulated over its earlier use, the absorption of technology becomes much easier, making the interaction much more efficient than the traditional methodology of teaching-learning (Comas-Quinn, 2011; Hughes, 2005). The instructor and the students have to adapt for teaching and learning, respectively, via the blended model. The use of technology for students is not new in today's world as it is frequently employed in K-12 education these days. However, a significant change in the role of instructors in the BL environment is observed wherein they have to adjust to the teaching-learning needs in terms of pedagogical, technological, and psychological acceptance. The transition of changing the norm becomes a difficult process, particularly challenging the seasoned instructors who seem to have limited knowledge and access to technology. Even if this is not an issue, instructors have expressed dissent over the student's ignorance of online course content. Also, it is evident that a blended form of learning demands for the instructors to have an actively enhanced role in creating teaching methods, time consumption, and a continual reflection on meaningful teaching methods (Ireland et al., 2009; Mccown, 2010; Ocak, 2011).

The instructor-student relationship prevents behavioural problems, contributes towards academic success, improves and aids in student's holistic development, which may have a long-lasting impact on the student. The traditional classroom teaching is pretty simple as the instructor's immediacy and immediacy behaviours are immensely important. The term immediacy here represents a psychological distance between the instructor and the student, involved in the communication. The verbal (praising, soliciting views, humour, self-disclosure, etc.) and nonverbal (eye contact, facial expressions, gestures, physical proximity, etc.) immediacy behaviours can make the psychological space between them open and comfortable (Swan, 2001). However, this relationship is bound to change under BL wherein it takes an individualized anchor form to support the students at the educational institution, home, and society. It may be in the form of one-on-one interactions with instructors, peer discussion slots, counselling sessions, off-site mentorships, etc., slotted in their timetable which makes them feel they have a support network around them. As the instructor becomes a mentor or a coach, indifferent to a mere knowledge provider in the case of face-to-face teaching, the learning becomes more customized and student-driven. The approach changes to bottom-up instead of top-down, which gives the opportunity to the instructor for strengthening and having a more meaningful relationship. It can provide insights into students' personal struggles and needs, which may empower the instructor to mentor and coach through obstacles with providing comfort to the student.

Self-directed learning is an active teaching-learning approach. Self-direction is a skill and needs to be inculcated in the students. In this learning methodology, the students are responsible for their own learning and instructors are mere facilitators. Some students are great at self-management while some of them are not. Unequal classroom results in fast and independent learning for successful students, while unsuccessful students fall quickly through the cracks and feel lost. Bhandari et al. (2020) reported that since it is a different methodology, the students should track their progress, identify the areas which require effort, and make efforts towards self-development. Alternatively, facilitators should know about the abilities that require mentoring. A combined effort by instructors and students alike will be helpful to make autonomous and lifelong learners.

To present the information in a more appealing and absorbable way to the students, it is important for the instructor to know about the learning preferences and formats of an individual student. In terms of learning, students may be classified into active and passive learners. Active learners engage actively with the course content, within themselves, and with the instructor during their interactions. On the other hand, passive learners are less involved in the process and interactions resulting during the same (Baba et al., 2020). BL is an active learning classroom that allows the students to outperform their peers, exceeding their own grade expectancy along with providing significant learning gains in comparison to the traditional learning approach (Wheeler, 2018). The use of technology in BL may also reduce the dependence of the student over the instructor. The student can go onto a web search and instantly find the required information from numerous sources on the internet. However, various factors affect the success of technology's use in BL such as familiarity with technology usage along with autonomous learning along with socioeconomic conditions. Other factors such as self-efficacy help in understanding student's computer literacy, explaining their relationship when controlled for other student characteristics and contextual background variables (Hatlevik et al., 2018). Neumeier et al. reported 68% participants out of 350 students finishing the course which was implemented due to a host of BL methodology aspects (Neumeier, 2005).

8.3 PEDAGOGY OF BLENDED LEARNING

The first and foremost pedagogical aspect in the successful implementation of BL mode would be to prepare the learners. It is quite likely that the strategy of learning is part offline and online. Moreover, each portion would be treated with a lesser emphasis in the opposite phase. For example, one topic that is addressed in online mode would get less focus in face-to-face mode and vice versa. All such dynamics may be completely new to the learners and would come under planning right at the beginning. The provision of orientation and rationale for inclusion in such a situation needs to be a part of the teaching plan. The instructor will then need to introduce it at all levels, including peer instructors, participating learners, and higher authorities of the institute. Letting the learners know about the freedom, benefits, and diversity of resources and how BL can reduce cognitive load would make it easier for instructors to include these activities as a part of the teaching-learning process.

The mere shifting from a face-to-face mode of instruction does not lead to meaningful blended instruction. A holistic approach is required to ensure that all the components of a learning experience fit together in an effective way. There is a need to design instructional delivery models to retain the instructors' authenticity in blended mode while creatively addressing the different reservations put forth by the stakeholders involved. The most important stakeholders in the teaching-learning process are the instructors and learners. The concerns regarding technological and philosophical aspects could be at the personal and/or institutional level. The instructional design of teaching-learning involves a step-by-step system to address needs, design, and development of content and the evaluation of the effectiveness of the teaching-learning experience (Gagne et al., 2004). The virtues of instructional design in the form of systematic arrangement to facilitate intentional learning, assimilating learning theories, and carrying out analysis based on assessments and feedback are very much important and relevant for the online component of BL. Besides many other instructional design models, the ADDIE model, appear to be comprehensible and congenial. The three major steps from the ADDIE model of the instructional model, i.e., design, development, and implementation, need to be revisited for the reconceptualization of the complete learning experience (Kurt, 2017).

The pivotal role that the ADDIE model can play in designing blended courses and introducing newer pedagogies was studied in the context of technical education (Yousif, 2016). The Potential of Open Educational Resources (OER) during the implementation phase of ADDIE model in BL courses with a significant effect on learning were reported in this study. The limitations due to non-coverage over all the five steps of the ADDIE model were reportedly related to the insufficient evaluation of the impact of these pedagogies on the students' performance. An instructional design model in a blended environment for a complex engineering scenario was implemented for two groups of students using face-to-face components including direct instruction and online components including multimedia, carefully designed slides, and various instructional activities (Dalal et al., 2017). It was reported that this blended mode has resulted in increasing student interest and motivation, team building character, assessment of students understanding, with an iterative design process.

The implementation phase of the ADDIE model in a blended pedagogy needs the instructor to avoid any bias toward technology and treat all tools along with technology sequencing during the conduct of the course as valuable components. The inherent belief that in-person learning is more effective would be a major hurdle in the path of successful BL mode. It is important that instructors develop an objective attitude toward the mode and tools used in the teaching-learning process. The decision of selection of the tools and technology that best support the performance objectives and context outcome should come as an output of the analysis phase. The components of the analysis phase of the ADDIE model when superimposed on a BL environment could create the optimal blend. A full range of learning technologies are available, and personalized options like mentoring and shadowing experts can also play an effective part in a BL experience. The 'Evaluation' phase stands as the last stage of the ADDIE model encompassing the what, how, why, and when of the deliverables were accomplished. A framework and self-assessment mechanism including all the exhaustive dimensions for the sustainable and scalable BL model so as to build the capacity of higher education institutions is provided by Lim, Cher Ping (Fong et al., 2014). In this study, the philosophical point of view, institutional policies and infrastructural support, and research-oriented approach were included as major pillars of the self-assessment mechanism.

The practices of mutual recognition of engineering qualifications are present in the form of various agreements (Dixit & Pathak, 2012). The first of this kind of agreement was the Washington Accord accepted worldwide with 21 countries holding full rights and 7 countries with provisional signatories. The role of the accord is to establish and set benchmarks for professional engineering education across the participating countries. The focus of these activities is the consideration of being learner-centric in view of outcome-based education. In this scenario, the BL effectiveness also needs to be investigated considering grades, course completion, retention, and effectiveness regarding the learner characteristics/background, design features, and outcomes. The role of design features for BL which affect the learning outcomes are reported by Kintu et al. (2017b). This study concluded the characteristics and background of learners as strong factors worth considering in the design of BL environments. The study also reported that the outcomes of blended mode are unaffected by age group under consideration and gender of the participant learners, while workload management was crucial for the learning outcomes.

The educational challenges faced during the learning experience affect the learning outcomes. It is, therefore, essential for the designing of content delivery to meet learning outcomes, rather than focus on the persuasion to include or use specific technologies that are least important. When the focus is not on the inclusion of specific technology, the choice of the adopted approach fulfils the learning outcomes. The survey-based planning done by taking learner preferences into account during the design and development of the course as this would assist the instructor in identifying the environments that the learners prefer or find most convenient.

With the discourse of the pedagogical aspect, the next thing is the appropriate media and content selection. The key terms are the media referring to vehicles that are used to deliver content such as printed material, podcasts, videos, animations, and so on. Some instructional media may be more appropriate than others in a given learning environment, but no single medium is inherently better or worse than any other. The selection of a particular medium certainly affects how the content is designed for taking maximum benefit of unique attributes of that specific medium. The effectiveness relies on content selection, management, and curation. The resources created as a part of a particular requirement and for a different set of learners when made available in some other context would appear as too hard, too easy, too impersonal, or too dry to other sets of learners. Thus, content curation also helps instructors protect their learners from the dump yard of false, outdated, and plain outlandish information. The harm is multi-fold when learners access such content, and therefore it becomes the responsibility of the instructor to provide access to factual, verifiable, most updated, and relevant course content to the learners.

Designing from scratch is better than redesign an existing course or curriculum. It is recommended to take a fresh start with blended design with an accentuation on needing a new perspective. The rework on a course that has existing constraints by the previous approach would be carried forward (Bonk & Graham, 2006). The instructor can go creating his/her own learning material. In the situations, if the external content is to be used, the best practice would be to cultivate the most appropriate external content with legal access from multiple sources, filtering it carefully so that the most relevant content is passed on to the learners, sharing the learning material at the right time, and adding value to that content with commentary, context, and correct flow. The evolution of the instructors' role and content designing in BL for one course evolved after multiple offerings under consideration was discussed by Danchak and Huguet (2004). The important rationale in this case study was using video streaming as a means to motivate the learners and provide them a road map rather than a mere tool of information delivery, setting the online content volume as per the learners appetite of learning and conducting specific question-based discussion sessions. The use of LMS for the implementation of BL has been explored by Erik Black (2007). The deliberations to be made before adopting the LMS as a means of blended mode are identified exploring the features of LMS, understanding their flexibility, and fact-finding about the suitability of LMS. While the very idea of LMS was originated to support instructors with different content areas, teaching philosophies and instructional designs, the LMS are getting identified with the stagnation of high-quality, flexible BL environments (Feldstein & Masson, 2006).

Ultimately, as an essence of the above discussion, it is always helpful as a part of planning to have a handy checklist keeping track of all important things that are done. The matrix of various aspects involved and their modalities for successful implementation of a BL environment are summarized in Table 8.2.

The quality and effectiveness of instruction determines the extent of learning, while media selection depends upon the instructional objective with an aim of improving learners' performance. The classification of media being asynchronous and synchronous certainly would help instructors in this situation. The media in which the level of interactivity required is minimal is more appropriate for the lower cognitive levels and could be addressed asynchronously, and when the level of interactivity required is high for the higher cognitive levels, a synchronous learning environment could be provided. The point to concur here is designing synergy between the components of a blended environment that will fit together so that the learner could link all the events in the learning experiences from each component of the blend to each other and could see these components working to reinforce and augment each other.

8.4 TOOLS AND TECHNOLOGY

One of the important factors that made e-learning popular in recent times is the Cloud computing (CC) technology. The development and maintenance of the localized onsite web-based e-learning system in institutions or enterprises is a very difficult and expensive task due to the lack of support and underlying infrastructure. However, technologies such as CC can be availed as a service through the internet, enabling to build more successful and effective educational environments (AlCattan, 2014; Pocatilu et al., 2010). The tools used in BL for online collaborative learning activities can be broadly classified into three categories, i.e., synchronized tools, LMS tools, and social networking tools, as shown in Figure 8.1 (Al-Samarraie & Saeed, 2018).

TABLE 8.2
The Feature Selection Matrix of Various Aspects Involved and Their Modalities for Successful Implementation of BL Environment

Sr. No.	Pedagogical Aspects	Technological Aspects	Content Based Aspects	Interpersonal Aspects
1	Identifying and catering needs of different learners.	Fast paced and cost-effective	Coverage of thorough curriculum	Responsive behaviour and active involvement from instructor and learner
2	Student-led discussions and feedback from instructors, and peers	Ease of accessibility, convenient and flexible	Availability of exhaustive internal content as per the course requirements	Good connect with the peers and focus on interactive discussion
3	Clear course design distributing course content in online and face-to-face modes	Ease of analytical features	Copyright free external content	Building the sense of community and no learner feels left out
4	Acknowledging the learners interest and motivation	Availability regarding out of the class activities like animations, virtual experiments, interaction with content	Careful content curation	Social presence

8.4.1 Synchronized Tools

A variety of synchronized web applications are available that can be used for delivering content and engaging students for collaborative work. Google Apps is the most popular and extensively used tools for content delivery and collaborative work and includes Gmail, Google Drive, Google Calendar, Google Classroom, Google Meet, and so on (Brown & Hocutt, 2015). Google offers these tools both in the free version (individual Google account, G Suite for Education) and as a subscription-based version (Google Workspace). One of the most important advantages of the Google Apps is that various tools can be integrated in the Google Apps portfolio, which offers a powerful framework that ranges from basic communication to high-end video streaming tools (Brown & Hocutt, 2015; Ferreira, 2014). A similar kind of subscription service is offered by Microsoft known as 'Office 365' as a part of the Microsoft Office product line. Apart from these generic productivity and collaboration tools which offer these services as a complete solution package, there are several free and subscription-based tools that cater to the specific collaborative and interactive learning requirements. Although a host of representative tools that can cater to these needs exist, we are discussing a few of them as a part of the upcoming section as given below:

- Padlet (https://padlet.com/): It is a sort of virtual "bulletin" board where students and instructors can collaborate, reflect, share digital content in a

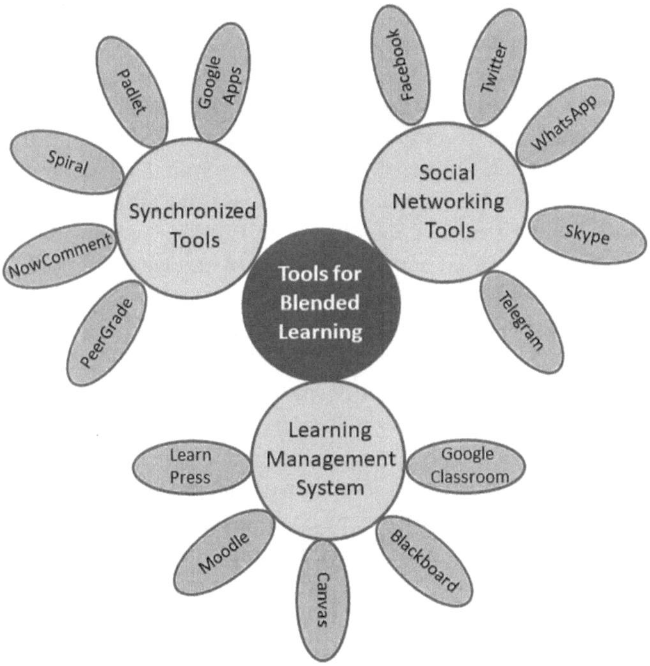

FIGURE 8.1 Tools for blended learning.

controlled environment. It offers free as well as premium services for individuals, educational institutes, and business users.

- Mural (https://www.mural.co/): It is a digital workspace that can be used by instructors and students to work together virtually to create, innovate, and solve important problems. Mural offers limited free services for schools and universities.
- Spiral (https://spiral.ac/): It is a free-to-use interactive learning platform which can be used by both instructors and students for conducting various flipped classroom activities such as quick formative assessment, student collaboration, and interactive videos.
- NowComment (https://nowcomment.com/): It offers free, easy-to-use, and engaging discussions over the online documents for students and instructors.
- PeerGrade (https://www.peergrade.io/): It is a free online tool wherein students can submit the assignments. PeerGrade automatically distributes solved assignments among the peer students. The students evaluate these assignments as per the rubrics set by the instructor and provide feedback.

8.4.2 Learning Management System Tools

An LMS is a software-based or software as a service (SaaS) platform that is designed to create, distribute, and manage the delivery of educational content, its assessment, and the evaluation of student learnings. The most popular form of LMS is a web-based interface that provides a wide range of facilities in the form of media resources, digital content, forums, questionnaires, chats, assignment, content sharing, and so on. There are several forms of available LMSs right from a very simple and limited delivery framework like LearnPress to complex environments with very rich features such as Moodle. Also, several open source LMS options are available like Moodle, ATutor, Open edX, Chamilo, Canvas, Forma, Dokeos, Ilias, and others (Cavus & Zabadi, 2014) along with commercial LMS options like Litmos, LearnUpon, TalentLMS, Blackboard, Google Classroom, etc. The decision of choosing the right LMS as per the requirement of BL is a critical task since each LMS has a host of features and certain possible limitations, for example, LearnPress LMS, a product of WordPress, is very cost-effective and easy to use and maintain. It is a web-based application used for delivering the online content in the form of lessons and quizzes, but it hardly provides any feature for student interaction like discussion forums or recourse sharing. On the other hand, Moodle is the most widely used open-source LMS with rich administrative tools, content creation, and delivery modules, assessment tools, communications tools, learning analytics, the scope for customization and integrations with other systems. However, the adaptation, implementation, and utilization of Moodle require special efforts in terms of infrastructure, finance, and training.

8.4.3 Social Networking Tools

Around 3.5 billion people globally make use of social networking sites (SNSs) or applications. The users spend 3 hours per day on average basis over these SNSs (https://www.oberlo.in/). The students are using SNSs as an alternative learning environment. Various social networking tools and messaging applications like Facebook, Twitter, Skype, WhatsApp, and Telegram are used by students and instructors for exchanging digital contents and informal learning (Ng et al., 2018). There are instances wherein tools like Facebook and WordPress were used in structured and informal setup for activities like online attendance, storage, and dissemination of course content, event scheduling for the lectures, and virtual meetings (Abdillah, 2016).

WhatsApp is the most popular free messenger application widely used among students to exchange multimedia messages like photos, videos, and audio along with simple text messages. The continuous availability of WhatsApp through smartphones has made it possible to learn anytime and anywhere (Gon & Rawekar, 2017). Zulkanain et al. applied the Community of Inquiry (CoI) framework to validate whether WhatsApp could be used for developing a pedagogical framework. Their findings reveal that it can indeed be used for educational purposes (Zulkanain et al., 2020).

8.5 CHALLENGES IN ADOPTION OF BLENDED LEARNING BY STAKEHOLDERS

BL approach is considered to be beneficial over face-to-face or completely online teaching-learning. The current situation due to the COVID-19 pandemic has created a sense of urgency to use this approach. However, adopting this approach is not as straightforward as it poses certain challenges from different stakeholders, i.e., students, instructors, and educational institutions.

8.5.1 Students

The major challenge faced by students is that they lack the self-motivation and regulation in the learning path; chances of distraction are more under the marginalized controlled learning environment. Some of the reasons for low levels of self-regulation and motivation include procrastination, lack of rigid timeframe, and limited interaction and socialization opportunities. The availability of social networking sites and other online distractions at the fingertips boost the procrastination behaviour. Some other challenges faced by students are related to the adaptation of new technology; students are not willing to invest their time in tools that have a steep learning curve. Also, the lack of student's competence and proficiency to use technology effectively for learning is a major challenge in the implementation of the BL approach. The constantly evolving technological variations add complexity, which makes it difficult to use the online component of BL. On many occasions, the students spend significantly more time on learning how to use these technologies rather than on the actual course content.

8.5.2 Instructors

The lack in the instructor regard is in the form of proficiency, time, and willingness to learn new techniques for teaching as it is an additional responsibility for the creation and delivery of the course content. The creation and maintenance of online content is a time-consuming and tiresome activity. The transfer from traditional teaching-learning to the blended mode requires a complete shift in teaching methodology, and many instructors, especially the seasoned ones, are reluctant to make such a massive change in their approach. The unwillingness to learn new constantly evolving technology is a major challenge posed to the instructors. Also, recording and editing videos for the online component of BL is one of the most challenging tasks for instructors. The increase in workload, at least for the initial implementation of blended mode, and adaptation to changing technology summarize some of the major challenges faced by the instructors in BL mode.

8.5.3 Educational Institutions

The lack of organizational aspirations, technology, and training for the overall implementation of BL are the challenges faced by educational institutions. The institutions face challenges in providing and maintaining the 'state-of-the-art' technologies required for successful implementation of the online component of blended mode. They lack the software and hardware support required to create and maintain an effective online learning environment essential in the BL mode. Moreover, adapting to evolving technology and developing new competencies at the institutional level is a big task as, on occasions, the policymakers in the institute lack the understanding about the BL approaches. Also, the institutions face problems in training the instructors and students alike regarding pedagogical and technological aspects of BL approach (Liu et al., 2020; Rasheed et al., 2020).

8.6 BLENDED LEARNING EFFECTIVENESS

The effectiveness of pedagogy and the intended benefits with possible concerns depends on many factors such as nature and the characteristic behaviour of the learners' cohort, the things learners are intended to learn, the knowledge they need to gain, and knowing how to use the learning and mediating processes associated. This process constitutes a broad range of components in curriculum and instruction that instructors orchestrate to promote learning, assessment, and evaluation. The intended benefits from a BL include but are not limited to:

1. Deeper, broader, and more flexible platform for knowledge transfer.
2. Learning experience outside the classroom and helping to create learning communities.
3. Potential of extending and enhancing the educational experience with learners accessing the content prior to face-to-face sessions
4. Focus of face-to-face session on "practical" or hands-on training
5. Platform to continue dialogue with peers and the instructor along with asking follow-up questions. Technical education demands the learners to learn most effectively when they get to apply knowledge in their own work environment and then get to reflect on that experience and ask questions to subject matter experts.
6. Extend the learning experience allowing learners more opportunity to apply, practice, and gain direct experience
7. Better learning experience and greater value overcoming the limitations of space-constrained learning, at the time when the presenter is available.
8. Learner-centred strategy where learners need to be provided with more options, time, and freedom to explore learning. The learning experience extends before or after the face-to-face interaction.
9. Availability of personalized content as per the needs of the learners and allowing learners to complete few of the activities independently at their own pace of learning.
10. Fetching the benefits of online examination, assessment, and evaluation.

The concerns that need to be overcome or possible disadvantages with a BL environment could be as follows:

1. Learners need to navigate between paper-based classroom activities and technology-based activities at different platforms.
2. Increase in screen time of the learner.
3. The availability of resources and technology may be challenging in certain situations.
4. Instructors workload gets increased in preparing the content and arranging the online mode activities.

5. Ensuring the competencies of the instructor as well as learners to cope with the use of technology, failing to which may create stigma and affect the offline part of teaching-learning process.

8.7 CASE STUDY

Alkhatib proposed a BL approach in the form of a flipped classroom for engineering curriculum. The course 'Structural Steel Design' from the undergraduate engineering curriculum (Bachelor of Science in architectural engineering program) was selected. This course was intended to teach about the design of structural elements in steel buildings, which include the design of steel tension members, beams, columns, and joints. Such a course requires technical skills, critical thinking, and problem-solving ability, which makes it difficult to offer as a completely online course. The interaction with instructors and peers plays an important role in the teaching-learning process, which makes it suitable for offering the course via BL mode. The course traditionally in face-to-face mode comprised of theory lectures, problem-solving sessions, and software training sessions. On the other hand, in flipped classroom approach, content delivery or instructional part was moved out of class with the help of technology as an online component, and structured instructor-student along with student-student interaction was carried out in the class as a face-to-face component. With this flipped classroom approach, students were supposed to prepare for class by using online resources to reduce lecture time, thus allowing more engagement in class during face-to-face sessions.

The teaching program for the 'Structural Steel Design' course was divided into two components as per the requirement of BL. The online component consisted of notes, problems, exercises, and recorded videos for each lecture that were made available to students well before the face-to-face session followed by in-class interactive activities. Furthermore, the online component was divided into pre-class and post-class activities. The interactive lecture module document was designed and comprised of brainstorming questions, short exercises, problems, and video lectures for the execution of pre-class activities. In the classroom or during face-to-face sessions, the students interacted with the instructor and peers on the exercises they tried before this session. An attempt was designed to engage the students in higher-level thinking activities like synthesis and analysis. An interactive presentation software 'SlideDog,' which is also equipped with an active chat window for communication between students and instructor, was utilized for managing the activities during face-to-face sessions. Students can answer the quiz through SlideDog on their smartphones and instructors can manage and share all these activities on SMART Board.

The student attainment levels were assessed before and after the implementation of the proposed interactive teaching method based on four different parameters such as student results, student opinion, instructor opinion, and program indicator. The higher values for student results were obtained after the implementation of the method and was attributed to the availability of online course material, enhanced students' interaction, and improved student preparation for face-to-face sessions. A survey was conducted to evaluate the student opinion about blended teaching mode, which revealed that the students preferred using the interactive learning environment over the traditional teaching method. A majority of students reflected that the pre-class online activities were helpful in enabling them to participate in the interactions during face-to-face sessions and made the lecture more interesting and comprehensible. The author reported the improvement in learners' performance with such a pedagogical approach from 3.9 to 4.4 on a scale of 1–5. Other reported benefits are ease in content retrieval and improved problem-solving skills, analytical, and critical thinking skills (Alkhatib, 2018).

8.8 CONCLUSIONS

There are enough evidences presented to establish the success of BL approach in teaching-learning. However, there are some essential components such as internet connectivity in developing and accessing any online content for the BL model. For students in rural areas with limited internet infrastructure and poor internet connectivity, it becomes an issue. However, even in urban areas where internet speed is good, e-learning is hampered by limited data availability and cost associated with its use. The recent pandemic due to COVID-19 revealed several challenges, especially internet connectivity, that may lead to ineffective implementation of BL mode. It was experienced that the platforms used for online content delivery are not compatible with the devices that the students use to access the online content. Most of the students depend on smartphones, making it difficult to access the platforms which are designed specifically for desktop. The students have reported the instructor's unpreparedness and lack of technical competence for delivering online content. Some learners had negative attitudes about the effectiveness of online learning mode along with reporting the effectiveness of face-to-face teaching upon comparing with 100% online teaching as it restricted the opportunities for socializing. The complete unavailability of face-to-face components in teaching-learning emphasizes the need for the instructional strategy to be blended into online as well as offline activities which further strengthens the case for using BL mode of teaching-learning. Simultaneously, educators need to embrace new approaches for redesigning the knowledge transfer processes and be more creative for the creation and delivery of educational content.

REFERENCES

Abdillah, L. A. (2016). Exploring students blended learning through social media. *ComTech: Computer, Mathematics and Engineering Applications,* 7(4), 245. https://doi.org/10.21512/comtech.v7i4.2495.

AlCattan, R. F. (2014). Integration of cloud computing and web2. 0 collaboration technologies in e-learning. *ArXiv Preprint ArXiv:1406.5020.*

Alkhatib, O. J. (2018). An interactive and blended learning model for engineering education. *Journal of Computers in Education, 5*(1), 19–48.

Allen, E., & Seaman, J. (2010). Class Differences: Online Education in the United States, 2010. 30.

Al-Samarraie, H., & Saeed, N. (2018). A systematic review of cloud computing tools for collaborative learning: Opportunities and challenges to the blended-learning environment. *Computers & Education, 124*, 77–91.

Archibald, D., & Worsley, S. (2019). The father of distance learning. *TechTrends, 63*(2), 100–101. https://doi.org/10.1007/s11528-019-00373-7.

Baba, K., Cheimanoff, N., & El Faddouli, N. (2020). A comparative study of active and passive learning approaches in hybrid learning, undergraduate, educational programs. In: Arai, K., Kapoor, S. & Bhatia, R. (Eds.). *Intelligent Computing* (pp. 715–725): Springer International Publishing. https://doi.org/10.1007/978-3-030-52249-0_48.

Berge, Z. L. (1999). Interaction in post-secondary web-based learning. *Educational Technology, 39*(1), 5–11.

Bernard, R. M., Borokhovski, E., Schmid, R. F., Tamim, R. M., & Abrami, P. C. (2014). A meta-analysis of blended learning and technology use in higher education: From the general to the applied. *Journal of Computing in Higher Education, 26*(1), 87–122.

Bhandari, B., Chopra, D., & Singh, K. (2020). Self-directed learning: Assessment of students' abilities and their perspective. *Advances in Physiology Education, 44*(3), 383–386. https://doi.org/10.1152/advan.00010.2020

Black, E.W. (2007). The other side of the LMS: Considering implementation and use in the adoption of an LMS in online and blended learning environments. *TechTrends, 51*(2), 35–39. https://doi.org/10.1007/s11528-007-0024-x.

Bonk, C. J., & Graham, C. R. (Eds.). (2006). *The Handbook of Blended Learning: Global Perspectives, Local Designs* (1st ed.). San Francisco, CA: Pfeiffer.

Brown, M. (2015). Mapping the Study of Blended Learning in Engineering Education. *2015 ASEE Annual Conference and Exposition Proceedings*, 26.1133.1–26.1133.15. https://doi.org/10.18260/p.24470.

Brown, M. E., & Hocutt, D. L. (2015). Learning to use, useful for learning: A usability study of Google apps for education. *Journal of Usability Studies, 10*(4), 160.

Caner, M. (2012). The definition of blended learning in higher education. In *Blended Learning Environments for Adults: Evaluations and Frameworks* (pp. 19–34). IGI Global. https://www.igi-global.com/chapter/content/65192.

Cavus, N., & Zabadi, T. (2014). A comparison of open source learning management systems. *Procedia-Social and Behavioral Sciences, 143*, 521–526.

Collis, B. & Moonen, J. (2001). *Flexible Learning in a Digital World: Experiences and Expectations*. Routledge & CRC Press. https://www.routledge.com/Flexible-Learning-in-a-Digital-World-Experiences-and-Expectations/ollis-Betty-Moonen-Jef/p/book/9780749433710.

Comas-Quinn, A. (2011). Learning to teach online or learning to become an online teacher: An exploration of teachers' experiences in a blended learning course. *ReCALL, 23*(3), 218–232. https://doi.org/10.1017/S0958344011000152.

Dalal, M., Larson, J., Savenye, W., Zapata, C., Hamdan, N., & Kavazanjian, E. (2017, April 21). Supporting Engineering Education with Instructional Design: The Case of an Introductory Module on Biogeotechnical Engineering.

Danchak, M. M., & Huguet, M.-P. (2004). Designing for the changing role of the instructor in blended learning. *IEEE Transactions on Professional Communication, 47*(3), 200–210. https://doi.org/10.1109/TPC.2004.833684.

Demetriadis, S., Barbas, A., Molohides, A., Palaigeorgiou, G., Psillos, D., Vlahavas, I., Tsoukalas, I., & Pombortsis, A. (2003). "Cultures in negotiation": Teachers' acceptance/resistance attitudes considering the infusion of technology into schools. *Computers & Education, 41*(1), 19–37. https://doi.org/10.1016/S0360-1315(03)00012-5.

Dixit, R. K., & Pathak, M. (2012). An overview of international engineering accords with special reference to washington accord. *Journal of Engineering, Science and Management Education, 5*, 467–471.

Driscoll, M. (2002). Blended learning: Let's get beyond the hype. *E-Learning, 3*, 2.

Dziuban, C., Graham, C. R., Moskal, P. D., Norberg, A., & Sicilia, N. (2018). Blended learning: The new normal and emerging technologies. *International Journal of Educational Technology in Higher Education, 15*(1), 3. https://doi.org/10.1186/s41239-017-0087-5.

Feldstein, M., & Masson, P. (2006). Unbolting the chairs: Making learning management systems more flexible. *ELearn, 2006*(1), 2. https://doi.org/10.1145/1125912.1125916.

Ferreira, J. M. M. (2014). Flipped classrooms: From concept to reality using Google Apps. *2014 11th International Conference on Remote Engineering and Virtual Instrumentation (REV)*, pp. 204–208. https://doi.org/10.1109/REV.2014.6784256

Fong, R. W., Lee, J. C., Chang, C., Zhang, Z., Ngai, A. C., & Lim, C. P. (2014). Digital teaching portfolio in higher education: Examining colleagues' perceptions to inform implementation strategies. *The Internet and Higher Education, 20*, 60–68. https://doi.org/10.1016/j.iheduc.2013.06.003

Gagne, R. M., Wager, W. W., Golas, K. C., Keller, J. M., & Russell, J. D. (2004). *Principles of Instructional Design* (5th ed.). Belmont, CA: Wadsworth Pub Co.

Ginaya, G., Rejeki, I. N. M., & Astuti, N. N. S. (2018). The effects of blended learning to students' speaking ability: A study of utilizing technology to strengthen the conventional instruction. *International Journal of Linguistics, Literature and Culture, 4*(3), 1–14.

Gon, S., & Rawekar, A. (2017). Effectivity of e-learning through WhatsApp as a teaching learning tool. *MVP Journal of Medical Science, 4*(1), 19–25.

Graham, C. R. (2013). Emerging practice and research in blended learning. In: Moore, M. G. (Ed.). *Handbook of Distance Education* (3rd ed.). London: Routledge.

Hatlevik, O. E., Throndsen, I., Loi, M., & Gudmundsdottir, G. B. (2018). Students' ICT self-efficacy and computer and information literacy: Determinants and relationships. *Computers & Education, 118*, 107–119. https://doi.org/10.1016/j.compedu.2017.11.011

Herbert, C., Velan, G. M., Pryor, W. M., & Kumar, R. K. (2017). A model for the use of blended learning in large group teaching sessions. *BMC Medical Education, 17*(1), 197. https://doi.org/10.1186/s12909-017-1057-2.

Hughes, J. (2005). The role of teacher knowledge and learning experiences in forming technology-integrated pedagogy—Learning & technology library (LearnTechLib). *Journal of Technology and Teacher Education, 13*(2), 277–302.

Ireland, J., Martindale, S., Johnson, N., Adams, D., & Eboh, W. (2009). Blended learning in education: Effects on knowledge and attitude. *British Journal of Nursing, 18*(2), 124–130.

Islam, S., Baharun, H., Muali, C., Ghufron, M. I., el Iq Bali, M., Wijaya, M., & Marzuki, I. (2018). To boost students' motivation and achievement through blended learning. *Journal of Physics: Conference Series, 1114*, 012046. https://doi.org/10.1088/1742-6596/1114/1/012046.

Jones, L. J. N., & Chew, E. (2015). *Blended Learning in Engineering Education: Curriculum Redesign and Development* (pp. 441–448). https://doi.org/10.1007/978-981-287-399-6_40

Kim, W. (2007). Towards a Definition and Methodology for Blended Learning. *International Workshop on Blended Learning 2007 (WBL).*

Kintu, M. J., Zhu, C., & Kagambe, E. (2017). Blended learning effectiveness: The relationship between student characteristics, design features and outcomes. *International Journal of Educational Technology in Higher Education, 14*(1), 7. https://doi.org/10.1186/s41239-017-0043-4

Kurt, S. (2017). ADDIE model: Instructional design. *Educational Technology*, 29.

Liu, Q., Geertshuis, S., & Grainger, R. (2020). Understanding academics' adoption of learning technologies: A systematic review. *Computers & Education, 151*, 103857. https://doi.org/10.1016/j.compedu.2020.103857.

Mccown, L. J. (2010). Blended courses: The best of online and traditional formats. *American Society for Clinical Laboratory Science, 23*(4), 205–211.

Nardi, B. A., Whittaker, S., & Bradner, E. (2000). Interaction and outeraction: Instant messaging in action. *Proceedings of the 2000 ACM Conference on Computer Supported Cooperative Work*, pp. 79–88.

Neumeier, P. (2005). A closer look at blended learning–parameters for designing a blended learning environment for language teaching and learning. *ReCALL: The Journal of EUROCALL, 17*(2), 163.

Ng, L. S., Thang, S. M., & Noor, N. M. (2018). The usage of social networking sites for informal learning: A comparative study between Malaysia students of different gender and age group. *International Journal of Computer-Assisted Language Learning and Teaching (IJCALLT), 8*(4), 76–88.

Norberg, A., Dziuban, C. D., & Moskal, P. D. (2011). A time-based blended learning model. *On the Horizon, 19*(3), 207–216. https://doi.org/10.1108/10748121111163913.

Ocak, M. A. (2011). Blend or not to blend: A study investigating faculty members perceptions of blended teaching. *World Journal on Educational Technology, 2*(3), 196–205.

Oliver, M., & Trigwell, K. (2005). Can 'Blended Learning' be redeemed? *E-Learning and Digital Media, 2*(1), 17–26. https://doi.org/10.2304/elea.2005.2.1.17.

Parker, A. (1999). Interaction in distance education: The critical conversation. *AACE Journal, 1*(12), 13–17.

Pocatilu, P., Alecu, F., & Vetrici, M. (2010). Measuring the efficiency of cloud computing for e-learning systems. *Wseas Transactions on Computers, 9*(1), 42–51.

Rasheed, R. A., Kamsin, A., & Abdullah, N. A. (2020). Challenges in the online component of blended learning: A systematic review. *Computers & Education, 144*, 103701. https://doi.org/10.1016/j.compedu.2019.103701.

Ross, B., & Gage, K. (2006). Global perspectives on blended learning: Insight from WebCT and our customers in highereducation. In: Bonk, C. J. & Graham, C. R. (Eds.). *Handbook of Blended Learning: Global Perspectives, Local Designs.* San Francisco, CA: Pfeiffer Publishing.

Singh, H., & Reed, C. (2001). *A White Paper: Achieving Success with Blended Learning.* 11.

Swan, K. (2001). Virtual interaction: Design factors affecting student satisfaction and perceived learning in asynchronous online courses. *Distance Education, 22*(2), 306–331. https://doi.org/10.1080/0158791010220208.

Ward, J., & LaBranche, G. A. (2003). Blended learning: The convergence of e-learning and meetings. *Franchising World, 35*(4), 22–24.

Watson, J. (2008). Blended learning: The convergence of online and face-to-face education. promising practices in online learning. In *North American Council for Online Learning.* North American Council for Online Learning. https://eric.ed.gov/?id=ED509636.

Wheeler, B. D. (2018). *Adopting Active Learning Classroom (ALC) Technology and Overcoming Barriers: A Faculty Development Intervention Model for Technology-Enhanced Learning Spaces.*

Yousif, A. A. (2016). The potentials of open educational resources to design blended learning courses with new pedagogies for significant learning. *Effectiveness of Bloom's Taxonomy In An Online Classroom-Improvements in Teaching and Learning Methods, 93*, 254–280.

Zulkanain, N. A., Miskon, S., & Syed Abdullah, N. (2020). An adapted pedagogical framework in utilizing WhatsApp for learning purpose. *Education and Information Technologies, 25*(4), 2811–2822. https://doi.org/10.1007/s10639-019-10096-0.

9 Shifting to Online Mode of Teaching during the COVID-19 Outbreak

Experiences of Faculties at a Higher Education Department in Pakistan

T. Noor and A. Nawab
Sukkur IBA University

CONTENTS

9.1 INTRODUCTION

The 21st century demands such complex skills not expected from a graduate of past. Griffin and Care (2014) organize these skills into four groups, namely, ways of thinking (creativity and innovation, critical thinking, problem-solving, decision-making, learning to learn, and metacognition), ways of working (communication and collaboration), tools for working (information literacy and ICT literacy), and living in the world (citizenship – local and global, life and career and personal and social responsibility – including cultural awareness and competence). Education in the 21st century ought to be focusing on these skills in order to enable the graduates to

DOI: 10.1201/9781003143796-9

succeed in the turbulent world (Ortega & Falconer, 2015). However, it is less likely to achieve these skills using the traditional face-to-face model of teaching and learning. Owing to many strengths such as scope for innovation, flexibility, and accessibility, online teaching and learning has the possibility to achieve or to support to the achievement of many of the 21st-century skills (Dhawan, 2020).

Consequently, referred to using various terms such as online learning, e-learning, web-based learning, blended learning, and so on, the ability to learn using computer connected to network has been around for several decades. Some scholars considered it as a need of time or necessity of future (Dhawan, 2020). However, organizations and individual had been showing reluctance to adopt the online mode of teaching and learning. What accelerated the process or transition to online teaching is the recent COVID-19 outbreak. The virus also taught us that things would have been very different today if we had seriously considered the online learning before the outbreak. We also realized the importance of 21st-century skills which, despite consistent reminding by scholars, we had been ignoring. Now online learning is no more an option. It has become a necessity. The organizations, teachers, and students who were previously reluctant to use ICT unwillingly shift to online mode for their very survival because they are now judged by their ability to adapt to the changes in such a short period. Initially, quality was not a concern. Just adopting online learning was what the institutions were concerned about (Carey, 2020). The International Association of Universities and UNESCO reported disruption to regular learning practices of over 1.5 billion higher education students in 185 countries due to university closures as a result of COVID-19 since April 2020 (Crawford et al., 2020). As the institutions lacked time for planning and development of online programs, the transition was full of issues and challenges.

Although we have sufficient understanding of the issues and challenges which the educational institutions and teachers faced in shifting to online mode in many other contexts, we still do not know the experience of teachers in the context of Pakistan, especially in the higher education system. Given that different contexts have their own realities, teachers in Pakistan might have experienced the shift in different way. This chapter reports the findings of a research conducted to explore the experience of faculties at a higher education department in Pakistan to understand the challenges they face and the strategies they used in relation to online teaching during the COVID-19 outbreak. In the following section, a brief analysis of literature on the issues and challenges related to online learning reported from other contexts is presented.

9.2 ISSUES AND CHALLENGES THE PANDEMIC BROUGHT

A systematic and careful analysis of the existing literature reveals a huge variety of issues and challenges institutions and individuals faced in shifting to online learning. The very first and major issue was related to resources, especially in the underdeveloped and developing regions of the world (Kalloo, Mitchell, & Kamalodeen, 2020; Qazi et al., 2020). Many institutions, teachers, and students lacked access to internet (Kalloo et al., 2020). The situation was exacerbated by the electricity-related issues prevalent in many developing countries (Dhawan, 2020). Students coming from poor background were unable to afford gadgets and internet connection. On the basis of such experiences, there was a risk of digital divide and, thus, achievement gaps created by the so-called shift to online mode of teaching and learning (Murphy, 2020b). Poor students who were hardly able to attend face-to-face classes remained deprived of education due to their remoteness and the lack of resources. Coupled to the resource factors, many technical issues (Almaiah, Al-Khasawneh, & Althunibat, 2020), such as downloading errors, issues with installation, login problems, and problems with audio and video, were also the factors significantly influencing the online teaching and learning processes during the pandemic.

Perhaps the most significant challenge was the skill required to not only use ICT but also engage students in meaningful learning activities (Agormedah, Henaku, Ayite, & Ansah, 2020). Teaching online is quite different from face-to-face teaching (Scull, Phillips, Sharma, & Garnier, 2020). Planning and implementing online teaching activities requires substantial skills that many of the teachers were lacking (Kalloo et al., 2020). Similarly, students also require skills to use ICT and be engaged in online learning (Aini, Budiarto, Putra, & Rahardja, 2020). Many of the students and teachers especially in the developing context were experiencing online teaching and learning for the first time. The efforts to transition were without preparation and training (Agormedah et al., 2020; Aini et al., 2020; Hoq, 2020) and, thus, the challenges they faced were very predictable. Some teachers even resisted the change as they were habitual to face-to-face teaching (Dhawan, 2020).

Another significant challenge that the teachers faced was engaging students and attaining their attention (Chen et al., 2020; Ortega & Falconer, 2015). Online learning was unexciting for many students given that they were used to learn through interaction, collaboration, and working together (Chen et al., 2020). Now listening to only teacher who lacked skills to engage them in interactive activities was monotonous for students. Some researchers even considered the technical skills secondary to social skills required for successful online learning (Scull et al., 2020). Some academics raised the question in relation to the difference between information and knowledge creation (Hoq, 2020). It was argued that using ICT, students would access information but the social aspect or the experiential learning or knowledge creation possible in face-to-face interaction would barely be achieved by online mode.

Some studies (for example, Kalloo et al., 2020) also reported the lack of parents' supervision as a hindrance in engaging students in active online learning. Historically, educational institutions have been considered having the sole responsibility in facilitating student learning. Exploiting home as a learning environment for students was a first-time experience for parents. The lack of supportive environment at home, the panic of the pandemic, isolation, loses of relatives, and so on also resulted in stress, fear, anxiety and depression, leading to the psychological or emotional problems consequently resulting in the lack of focus and concentration on studies (Chen et al., 2020; Dhawan, 2020).

9.3 STRATEGIES TO ADDRESS THE CHALLENGES

Although the pandemic brought many challenges, educational institutions and teachers gradually adapted to those challenges. The foremost significant factor that helped the institution in successful transition to the online mode was the availability of digitally confident teachers (Hoq, 2020). The pandemic demanded innovative strategies to engage students in learning with the help of ICT. Planning, designing, and implementing innovative instructional materials; engaging students in learning activities; and assessing their progress were such roles that were quite unlikely to be carried out by teachers lacking digital competency. Against this background, support mechanism (Dhawan, 2020; Hoq, 2020; Murphy, 2020a) provided to both the teachers and students in developing their digital literacy skills were highly recommended and helpful in successful shift to e-learning.

Studies suggested that to engage students in active online learning, the learning materials should be dynamic, interesting, and interactive (Dhawan, 2020; Ortega & Falconer, 2015). More collaborative, group-based, case learning and project-based learning proposed by earlier academics (Kim & Bonk, 2006) were further reiterated to engage students in active and meaningful online learning. An interesting innovative strategy was the concept of virtual world introduced in the Education Innovation Centre, University of the West England, Bristol, UK (Ortega & Falconer, 2015). The authors used a three-dimensional virtual space to provide postgraduate students with the opportunity to interact with each other and with the world through avatars aimed to develop a sense of community and connectedness. Consequently, creativity as the ability for imaginative responses was more helpful in facilitation of effective learning during the crisis period (Kalloo et al., 2020).

Some other strategies to engage students and to keep them on track were recording lectures, setting time limits for various tasks, developing stronger social connection with students, giving clear instructions, seeking feedback from the students, and regularly showing their presence to give students an impression that teachers are always available for support (Dhawan, 2020; Keeton, 2004; Scull et al., 2020). Although the impact of the pandemic on student achievement is yet to be measured, surveys suggested that the satisfaction of students depended on the quality of service, platform, and interaction (Chen et al., 2020). Similarly, the accessibility and ease in use were also found to be satisfying factors for students experiencing online learning (Qazi et al., 2020).

The literature analyzed above shows that substantial knowledge has been created in relation to the challenges the pandemic brought and the strategies used to address the emerging challenges in the context of education. However, our understanding of the emerging phenomenon is limited in many ways. First, as also suggested in earlier studies (see, for example, Chen et al., 2020), we still lack knowledge regarding the experience of multiple subjects especially teachers in different contexts in relation to the transition to online mode of teaching and learning. Second, although various educational institutions in Pakistan have shifted to online mode of teaching, their experience of the shift has not been studied and documented through a systematic research. In the context of Pakistan, we do not know how effective was the transition to online mode, what challenges the teachers faced, and how they addressed those challenges. We may not depend on the knowledge coming from other contexts with the fact different contexts have their own realities that bring different challenges and implications for the practices of teachers. Third, while almost all the educational institutions in Pakistan were expected to use one of the other online mode of teaching, the shift was unavoidable to higher education institutions owing to their level, availability of resources, and capacity of both teachers and students. Against this background, it was highly significant to study the experience of teachers in a higher education system. Researching their experience not only adds to the existing knowledge on the teachers' approaches in Pakistan during pandemic but also brings forth significant insights in relation to challenges and strategies peculiar to this region while shifting to online mode of teaching and learning. These insights will be highly helpful for policymakers, planners, teachers, and students in successful transition to online mode of education that has become a necessity of the current era.

9.4 METHODS

Aimed to explore the experiences of the faculties in shifting to online mode with the focus on the challenges they faced and the lessons they learned in making their online teaching effective, a qualitative research approach was used with the understanding that this approach will allow an in-depth exploration of the phenomenon by personally interacting with a wide range of participants in a natural setting and listening to their views (Bogdan & Biklen, 1998; Hancock, Okleford, & Windridge, 2007). Qualitative approach also allowed the researcher to further narrow down to one specific case using the case study method in order to capture a holistic view of the phenomenon (Stake, 2000). Although more than one department and larger number of teachers could have been included in the research, the intention to go deeper and to uncover all the significant factors related to the phenomenon was instrumental in narrowing the study to only one department. While the purpose is not to generalize the findings to other departments or higher education institutions in the context, given that these institutions represent the same system and share many similarities, there may be points for reflections for other higher education institutions and departments.

The department was selected conveniently as both the authors were located at this department at the time of data collection. This convenient selection was helpful to get easy entry, develop rapport, and collect data quite smoothly. In order to include a wide range of views representing diverse experience, subjects, and gender, ten faculties out of 23 were considered using purposive sampling procedure (Bloor, 2001) and contacted via email requesting their participation in the research. The email explained the purpose of the research, procedure of data collection, expectations from the participants, and other ethical considerations. Seven of the contacted faculties responded positively, out of whom five were taken as final samples as detailed in Table 9.1.

TABLE 9.1
Details of Research Participants

Position	Gender	Number
Assistant Professor	Male	2
	Female	1
Lecturer	Male	2
Total		5

To develop an in-depth understanding of the phenomenon in line with the purpose of our research, semi-structured interview procedure was preferred as this tool enables to use some guided questions, generate discussion, and probe deeper based on the responses of participants (Bloor, 2001). The interviews were audio-recorded with the permission of the participants. Once transcribed, the interviews were fed into NVIVO and coded as shown in Figure 9.1.

The codes were read again and again to identity patterns. The process of naming and renaming the codes and assigning them to different patterns continued till the patterns were finalized. As a result of the detailed process of coding, identifying, and modifying patterns, there emerged four major themes, namely

a. Dilemma of Learning Versus Assessment
b. Maintaining Students' Motivation, Attention, and Engagement
c. Connectivity and Technical Factors
d. The Strategies and Opportunities Faculties Exploited

9.5 RESULTS

In the following sections, data are presented and interpreted under the four themes listed above.

9.5.1 Dilemma of Learning versus Assessment

One of the major challenges hindering the smooth transition to online teaching as highlighted by the participants was the prevailing approach to learning and assessment. As revealed by the faculties, historically students have been showing reduced interest toward the process of learning mainly because of the systemic expectations. The system and resultantly students, look for something tangible or an evidence to prove that learning happens. Such evidence is mainly associated with marks obtained as result of attempting test or an assigned task. As Faculty A stated, "the system requires evidence that we are teaching, and students are learning. At times it looked the process was controlled, and the pedagogue himself or herself becomes vulnerable". As this representative quote suggests, the process is in a way predetermined where the teachers are expected to follow the common trend with limited possibility to assign innovative tasks that carry no score and thus no value for students.

While further elaborating this issue, Faculty C, another faculty opined that "shifting to online mode means engaging students in activities which are more self-directed, or students centered where the students are expected to carry out some projects or assignments to develop a deeper understanding of the course contents". However, as he further stated, when students realize that such activities carry no scores, they show reluctance to attempt them. Online learning is more about self-directed learning for which the students have not developed proper attitude and skill working in the current system, and this reinforces the need and importance of scores.

These practices restrict the possibilities of engaging students in various creative and innovative activities. Even if faculties want to do something innovative, given the existing approaches to learning and assessment, they show reluctance. As Faculty

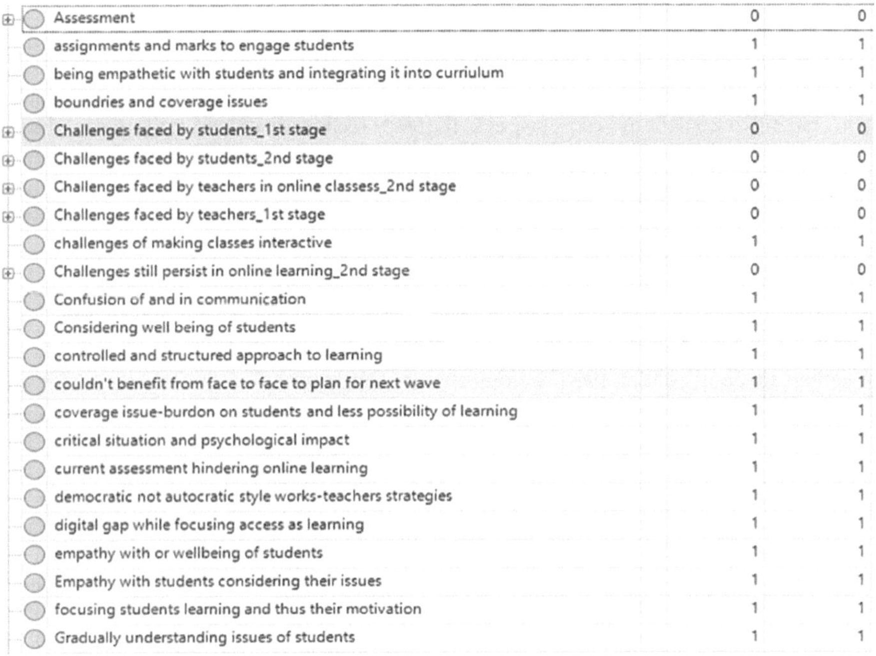

Assessment	0	0
assignments and marks to engage students	1	1
being empathetic with students and integrating it into curriulum	1	1
boundries and coverage issues	1	1
Challenges faced by students_1st stage	0	0
Challenges faced by students_2nd stage	0	0
Challenges faced by teachers in online classess_2nd stage	0	0
Challenges faced by teachers_1st stage	0	0
challenges of making classes interactive	1	1
Challenges still persist in online learning_2nd stage	0	0
Confusion of and in communication	1	1
Considering well being of students	1	1
controlled and structured approach to learning	1	1
couldn't benefit from face to face to plan for next wave	1	1
coverage issue-burdon on students and less possibility of learning	1	1
critical situation and psychological impact	1	1
current assessment hindering online learning	1	1
democratic not autocratic style works-teachers strategies	1	1
digital gap while focusing access as learning	1	1
empathy with or wellbeing of students	1	1
Empathy with students considering their issues	1	1
focusing students learning and thus their motivation	1	1
Gradually understanding issues of students	1	1

FIGURE 9.1 NVIVO coding.

A stated, "I have many ideas and possibilities in my mind but when I look at assessment demands, I curve on that. If I were independent, I would have shifted students to self-directed learning or projects." As this quote suggests, flexibility in the system will allow the faculties to use a variety of innovative strategies to engage students in online learning. Building on these experiences, the faculties were of the view that all the stakeholders require to conceptualize their practices in relation to teaching, learning, and assessment. They were of the view that our rigid and structured approach hinders achievement of the 21st-century skills. As Faculty C opined:

> Today education or literacy should be more about analytical skill, critical thinking skill, creative skill, evaluative skill, and reflective skill. I cannot say the content, or the scores are less important. But more important are the skills which we fail to focus in our teaching. So, all the stakeholders from policy makers to teachers should have the orientation that grading, or content is not important. Learning and development is important and that should be the focus.

Consequently, for the successful online teaching and learning, the faculties suggest reconceptualizing curricula with a focus on the learning processes and flexibility in the use of pedagogies instead of following predetermined strict criteria that associate learning to only marks. As the faculties believed, while shifting to online mode, students have been provided with access or platform. But for the interviewed faculties, access is not learning or education. It is a step or door toward learning.

9.5.2 Maintaining Students' Motivation, Attention, and Engagement

In the journey of shifting to online mode of teaching, another significant challenge faculty faced was maintaining students' motivation, attention, and engagement. Faculties considered these aspects the very basics for unless, and unless, these aspects were achieved, ensuring that students' learning was very unlikely. As Faculty B opined, "students will easily overcome all other issues if they are motivated. What challenged us in the very outset in our online journey was the low motivation level of students and the challenge to maintain their attention". As the faculties revealed, there were several reasons influencing the level of attention and motivation of students and consequently affecting their active engagement in the online learning. The first and foremost factor was the systemic approach to learning and assessment as presented in detail under the previous theme. Since many of the online activities such as Zoom sessions and some other assignments need not necessarily carry scores according to the expectation of students, they were showing reduced interest toward the online activities. Students, on the other hand, were trying to put most of their energy on the assignments or quizzes that carried scores. Consequently, it was challenging for the faculties to engage students in such activities which, although very significant for their learning, carried no or low score.

The second challenge was because of the physical remoteness. Although the faculties were trying to reach out to students using various platforms such as Zoom, it was challenging to ensure whether students are there and attentive. As Faculty D reported, a student may just sign in to Zoom for attendance purposes but he or she may not listen to our lecture and be busy in some other activity. We do not see them. Faculty A supported this view stating that "the scenario is beyond control. For example, Ayesha (a hypothetical name for a student) is represented by her microphone there. Microphone representing the student. 'Ayesha' is written there. There is total I can say absenteeism in engagement. You just need to imagine that students are there". In this background, faculties were trying to engage them using many strategies. However, such strategies were also not very much helpful due to certain reasons. For example, as shared by Faculty D, one strategy was to ask them turn on their video or mic to ensure they are attentive. But this strategy was not working for two reasons. First, female students were reluctant to turn on their video and the faculties were respecting this as a cultural and contextual norm. Second, many students had space issue in their house. When the mic or video was turned on, there used to be disturbance due to the background noise because of the presence of other siblings in the same room. The second strategy that the faculties tried to use was naming the students and asking them some questions in relation to the points covered in the lecture. This strategy was working with small groups. However, as Faculty E stated, "some groups consisted of huge number of students where it was very unlikely to name the students in order to maintain their attention". Some faculties also highlighted the issue peculiar to female students at home. During the face-to-face sessions, they were either in the hostels or coming to campus on daily basis. While at home, they were performing many household chores as expected of females in the context.

Faculties considered a big difference between online and face-to-face teaching in relation to the engagement of students. In the classroom, according to the participants, it is more convenient to engage students such as through eye contact, body language, asking questions, and active monitoring. Similarly, there was greater possibility to engage students in discussion and other collaborative group work. Since these strategies were not possible to be used in online teaching, maintaining the attention of students and engaging them in active learning was quite challenging. Consequently, the physical distance brought many challenges for faculties in relation to engaging and maintaining students' attention. Faculties interviewed for this research collectively believed that the physical remoteness was a hindrance in active learning. Due to the physical remoteness, they were unable to ensure that the students are attentive and learning was taking place.

9.5.3 Connectivity and Technical Factors

Connectivity and technical factors were also identified, significantly influencing the shift to online mode of teaching and learning. Connectivity was the major issue highlighted by all the participants interviewed for this research. Although some of the faculties also faced connectivity issue mainly because of their remote locations, the issue was more prominent to students. To begin with, many of the students lacked the very ICT devices, especially laptops, to connect to internet and to do their assignments. Faculties shared such examples where students either lacked any device or one laptop was used by many siblings for online learning in a family. As Faculty B stated:

> Some of them had no electronic gadgets, one of the students told me that he even does not have a smartphone. And one of the students shared with me that they have one laptop at home. And then there are four siblings who have to attend the online classes. So, they were waiting for their turns that who will use this laptop and at what time.

Although majority of the students were having devices, many of them lacked access to internet. These students usually come from remote areas where internet facility has not yet been provided. Some have poor signal issues. Faculty E intimated that students attend the session but get disconnected many times during the session. As he further stated, "I understand that it is because of internet connection; sometimes it is better, and sometimes it is bad". Faculties also shared examples where students were going to distant areas in search of the signals. One student requested to a faculty to wait for a while as traffic was passing. In search of signals, the student had shifted to such an area.

Coupled with the issues related to the availability of IT devices and internet, affording the internet packages was also a challenge for students. Those who do not access Wi-Fi facility – many of the students had this issue – used to purchase internet data from mobile services. Since students were supposed to be online most of the time to attend session or to perform the Learning Management System (LMS)-related tasks, it was quite costly for them to afford sufficient internet data. Similarly, power or electricity was also an issue for many students. As Faculty C revealed, "sometimes, students got disconnected and returned after long time when the session was almost closing. Students missed a lot due to such disruption of electricity".

Where the connectivity issues in the form of access to devices, internet, and electricity influenced the process of online learning, limited IT skills or using ICT for teaching and learning purposes was another significant challenge for the faculties. As Faculty E reported, "it was a sudden and new experience for us. Using LMS, Zoom or other platforms without any proper training was something we were not ready for". As Faculty A depicted the scene:

> There is learning in mobile, but it requires high expertise, I feel that as pedagogue has to learn teaching styles to teach using ICT, learners also need to learn the skills or ways. It is different than having a conversation in the class. Proper use of ICT such as engaging students in self-directed learning and project-based learning could have yielded more. However, how to design such projects online and how to engage students in such activities? I realize I need to develop my technology skills to successfully design and implement such online tasks.

As the excerpt suggests, teaching online requires skills different from face-to-face teaching. While face-to-face teaching requires training, skill development, and preparation, online teaching requires greater capacity and skills, being more complex than face-to-face teaching. In addition, faculties also referred to many other technical issues such as uploading materials, registering to various platforms, and using other options on Zoom and LMS. Without any training, a sudden shift to online teaching was, therefore, a huge challenge for the faculties.

9.5.4 The Strategies and Opportunities Faculties Exploited

While teaching online, although the faculties faced many challenges, they also came up with a variety of strategies to overcome those challenges. The foremost strategy that majority of the interviewed faculties used was considering the psychological problems of the students, being empathetic with them, and providing emotional and psychological support. As Faculty E mentioned, "our first priority was students' well-being. Due to pandemic, everyone was psychologically under stress. So asking students and acknowledging their issues was our main strategy to make them feel comfortable. We were trusting what they were saying as we were understanding their challenges". Thus, the first significant step that faculties carried out was to be empathetic with students so that they could feel comfortable and be mentally prepared for the shift.

Once students were on board, the challenge of engaging them was addressed by using various forums such as LMS and creating discussion forum. As Faculty D reported, "during online teaching we did not know whether students are reading the suggested material or not. So we posed questions to students related to the reading material". Students were, then, expected to post the responses in the discussion forum created on LMS. This strategy, as the faculties believed, proved very helpful to engage students. Similarly, to make sure students were attentive to the lectures, the faculties were asking questions randomly from students. As students were expecting a question from the teacher any time during the lecture, they were carefully listening to the teacher.

Flexibility was another significant lesson and strategy that the faculties learnt during their practice to make the online teaching effective. As the faculties collectively reported, flexibility was in the form of allowing different platforms, using variety of assessment strategies, and being flexible with deadlines. When some of the students were unable to access Zoom or LMS at times due to connectivity issues, they were allowed to use other applications such as WhatsApp. Faculties also recorded their sessions and uploaded it for students to access any time. Similarly, students were allowed to contribute to the discussion forum on LMS on the basis of their convenience. Considering the technology issue, some students were sent instructions via mobile message and allowed to write their assignments on paper and then to send them in hard copy form. All the above examples suggest that, on the basis of the emerging issues, teachers used a variety of strategies for a smooth transition and to make the online teaching and learning effective.

Although the faculties used many helpful strategies to make the online learning effective, a deeper and critical analysis of data showed several opportunities available to teachers that could have been exploited in a more effective way. One example is the availability of LMS platform at the institution. According to the faculties, had this platform been utilized before pandemic for teaching and learning purposes, the transition to online mode during the COVID-19 pandemic would have been smoother. As Faculty A stated, "although, teachers and students had already been provided with ICT facilities such as LMS years before this emergency, but unfortunately, we were not using it properly. Thus, in online classes most of the teachers and students were unaware of how to use it". This example shows a lack of taking proactive measures and exploiting the existing opportunities.

A similar attitude was reported when teachers got an opportunity of face-to-face sessions when the classes resumed after the first wave of COVID-19. The pandemic was still around. A proactive measure would have been exploiting this opportunity and planning to systematically address the issues and challenges faced during the first wave. The faculties were optimistic that the face-to-face sessions will continue and the pandemic will not hit again. As Faculty C intimated, "even the management was optimistic like us and didn't take any measure to prepare faculties for another wave or the pandemic. May be we received a couple of emails to develop course outline by keeping such emergencies in mind, however, there was not stress or other preparation."

Then hit the second wave of the pandemic and the classes were again shifted to online mode. Although this shift was relatively smooth, still there were such issues that could have been avoided by preparation during the face-to-face interaction. Thus, the data show a lack of proactive measures by both the management and faculties.

9.6 THE EMERGING INSIGHTS

Some of the results emerging from this research may not be surprising given that such factors influencing the shift to online mode of teaching have already been reported from other contexts. These factors, for example, include the connectivity, skill, and technical-related aspects (Almaiah et al., 2020; Dhawan, 2020; Kalloo et al., 2020; Qazi et al., 2020). As the faculties interviewed for this research indicated, these aspects highly influenced the smooth shift to online teaching. Although the majority of teachers had no issues with regard to access and connectivity, many of the students faced this problem as they lacked either devices or internet facility, being located in remote rural areas. Since online teaching and learning demanded a different approach in relation to planning and implementing teaching and assessment materials, some of the teachers and students, due to their lack of skills, struggled to effectively use online mode for teaching and learning purposes. This apparently simple findings have tremendous implications for the education system, especially in underdeveloped regions and countries. The views of faculties supported earlier studies (see for example, Murphy, 2020b) that the lack of access and connectivity will widen academic gaps between those who access online learning and those who lack this facility. The apparent agenda of those who propagate e-learning is to make education accessible to anyone irrespective of their location. However, given that some learners especially in backward contexts lack access and connectivity, depending on e-learning for education in those contexts is depriving them from the basic right of education. The importance and need of online learning may not be denied. However, our shift to online mode should not create further gaps in the already prevailing socioeconomic status, mainly prevailing because of the lack of equal access to education. Therefore, before shifting to online mode of teaching and learning, it must be ensured that students, especially those in remote and backward regions, have adequate access to resources required for online learning.

The second significant insight emerging from this research and less reported in existing literature is the learning versus assessment dilemma of shifting to online mode. As the faculties interviewed for this research believed, our current education system is characterized by such assessment practices where scores are preferred over learning. The education system demands something tangible such as a written response to a given question and then score allotted for the tangible output. In a way, education system depends on predetermined indicators that a learner should follow to demonstrate that learning has happened. This approach to learning and assessment highly influences the process of online teaching and learning. Online teaching requires a teacher to design and implement a variety of innovative activities to engage students in active learning (Kalloo et al., 2020). Such activities demand self-directed learning and performing tasks that need not necessarily carry scores for their accomplishments. However, working in the existing system, the students have developed an appreciation for scores and attitude to work for incentives, making it challenging for a teacher to engage them in activities that do not carry any score. It is suggested that for a successful and smooth transition to online learning, we should reconsider our approach to learning and assessment. If we stress learning instead of scores, our students may develop interest in learning instead of scores. For this shift to happen, our whole curricula require a thorough reconceptualization.

The next emerging insight that merits discussion is the possibilities and challenges of achieving 21st-century skills using the online mode of teaching. Although the faculties highlighted many benefits associated with online teaching, they also questioned this mode owing to its limited possibility to facilitate interaction opportunities for students (Chen et al., 2020). One of the highly required 21st-century skills is social skills and the attitude to work in and appreciate diversity (Griffin & Care, 2014). These skills and attitude are more likely to develop if learners frequently interact both inside and outside the classroom. Face-to-face mode of education has the possibility to engage students in such activities that involve interaction among students. Some may consider a social gathering at university café as a great learning and socialization opportunity. Given this logic, online learning deprives students of many such physical interactions deemed necessary to develop social skill, a skill much required in today's global and pluralistic society. In addition, as the faculties highlighted, online learning confines learners to home and isolation which may lead to psychological and emotional issues (Chen et al., 2020; Dhawan, 2020) that could have been avoided if in the company of others. Consequently, online learning may not be an alternative to some of those opportunities created by face-to-face teaching.

Another interesting but appealing comprehension emerging from the views of research participants is that individuals and organizations in this context lack proactive measures to face any emergency situation. Although COVID-19 was a sudden outbreak, ICT and its use in education has been around for a couple of decades. In addition, academics had been highlighting the importance of e-learning, and some even considered it as a need of the present and future (Todorova & Bjorn-Andersen, 2011). Despite such persistent calls in this direction, the individuals and organizations in this region showed reduced interest to integrate ICT in education. Although higher education institutions were equipped with ICT resources and connectivity, no serious efforts had been carried out either from the management or faculties to systematically integrate ICT for teaching and learning purposes. One example could be the availability of Learning Management System (LMS) in universities. However, even this otherwise effective and readily available platform was rarely used by the faculties. Another example showing our failure to avail the available opportunities is the way we utilized the face-to-face time when the educational institutions opened at the end of the first wave of COVID-19. Although the pandemic was still around, there was hardly any realization that teaching would again be shifted to online mode. The prudent way would have been to systematically reflect on the online teaching and learning experiences during the COVID-19 outbreak and to be prepared for any such situation in future. When the second wave of pandemic hit, teaching was again shifted to online mode. Both teachers and students encountered many of the same challenges faced during the first wave as no preparation had been done to tackle those issues. This finding suggests that our organizations and individuals should develop a proactive attitude to encounter the similar and related emergencies in future.

This research also highlights the importance of self-directed learning for both students and teachers of the present and future era (Griffin & Care, 2014). Although the academics had consistently stressed the need of the self-directed learning as a significant skill of individuals in the 21st century, this approach had hardly received considerable recognition mainly because of our attitude of dependency. Had we given attention to and learnt this skill, many of the issues that we as teachers and students faced during the corona outbreak could have been avoided. The sudden shift to online mode of teaching and learning without any preparation and support left the teachers and students to their own efforts and trials. Those who were used to independent learning found it quite easy to learn and practice innovative strategies. A simple example is teachers' own efforts to trial with ICT skills including using LMS, Zoom, and so many other synchronized and asynchronized modes of teaching and learning.

9.7 CONCLUSION AND IMPLICATIONS

This study showed that due to their socio-economic conditions and remoteness, many students were not able to benefit from online mode of learning. This conclusion has implications for donor agencies, governments, policymakers, and educational managers. If made accessible to all, online learning will bring equity in education. Otherwise, the shift will further widen the already existing gap in education due to socio-economic conditions of learners.

This research also highlights the need of reconceptualizing our curricula for a successful and smooth shift to e-learning. The start should be with assessment because the assessment practices are said to drive the teaching practices. Instead of only summative or written tests, there should be ongoing and formative assessments where rewards are given for students' projects, creative work, collaborative inquiry, presentations, and other innovative tasks. Once such activities are rewarded, students will take active part in them, ultimately leading to their enhanced learning. However, teachers' capacity in designing, implementing, and assessing those innovative tasks will be a big challenge. The system should not only revise their assessment system but also support faculties to enhance their skills in relation to the intended shift. Organizing workshops, designing and implementing innovative assessments, and synergizing on the existing human resources in the higher education institutions would be positive and successful steps toward the shift. As a result, faculties will be able to engage students in self-directed learning activities using a variety of online sources and platforms. Such innovative activities such as projects, inquiry tasks, presentations, and so on may also facilitate collaboration among students. This will address the genuine concerns of some academics that e-learning deprives students from collaboration and social development.

The last and important message derived from this study is that organizations and individuals should take proactive measures to survive in this turbulent world. The winner of tomorrow will be those who pass through a rigorous and continuous process of learning and improvement (Finger & Brand, 1999; Lam, Chan, Pan, & Wei, 2003). Today, both individuals and organizations have huge opportunities to be exploited and to benefit from. E-learning was there before COVID-19 and was an opportunity to use and to integrate in our education system. Since we lack proactive attitude for learning and improvement, we missed the opportunity. The result was our reduced preparation for the emergency in the form of COVID-19 outbreak. This pandemic reinforces the lesson that we, as individuals and organizations, be proactive, avail opportunities and learn continuously if we have to keep abreast with the speedy process of change.

Thus, this research not only supports findings of many earlier studies but also add several new insights to the existing knowledge on e-learning with a focus on COVID-19 outbreak. It supports the widespread concern and fear that e-learning will widen the educational gap if the realities of students coming from low socio-economic background are not considered. Similarly, while debating the shift to online mode, academics and practitioners mainly focus on resource and skills. The unique contribution of this study is that it calls for rethinking our approaches to learning and assessment to make online learning a success. No matter how resourceful and skillful we are, it is very unlikely for students to fully benefit from online mode unless they prefer learning over score. Second, the study shows that both individuals and organizations in this region lack proactive measures to cope up with any emergency situation. A culture of dependency has also been restricting individuals to exploit the existing opportunities. These insights have implications for not only online learning but also our survival in this turbulent world.

REFERENCES

Agormedah, E. K., Henaku, E. A., Ayite, D. M. K., & Ansah, E. A. (2020). Online learning in higher education during COVID-19 pandemic: A case of Ghana. *Journal of Educational Technology and Online Learning, 3*(3), 183–210.

Aini, Q., Budiarto, M., Putra, P. O. H., & Rahardja, U. (2020). Exploring e-learning challenges during the global COVID-19 pandemic: A review. *Jurnal Sistem Informasi, 16*(-2), 57–65.

Almaiah, M. A., Al-Khasawneh, A., & Althunibat, A. (2020). Exploring the critical challenges and factors influencing the E-learning system usage during COVID-19 pandemic. *Education and Information Technologies,* 25, 5261–5280.

Bloor, M. (2001). *Focus Groups in Social Research.* London, UK: Sage.

Bogdan, R. C., & Biklen, S. K. (1998). *Foundations of Qualitative Research in Education: An Introduction to Theory and Methods.* London: Allyn and Bacon.

Carey, K. (2020). Everybody Ready for the Big Migration to Online College. *Actually, No. Disponível em:. Acesso em, 8.*

Chen, T., Peng, L., Yin, X., Rong, J., Yang, J., & Cong, G. (2020). *Analysis of user satisfaction with online education platforms in China during the COVID-19 pandemic.* Paper presented at the Healthcare.

Crawford, J., Butler-Henderson, K., Rudolph, J., Malkawi, B., Glowatz, M., Burton, R., … Lam, S. (2020). COVID-19: 20 countries' higher education intra-period digital pedagogy responses. *Journal of Applied Learning & Teaching, 3*(1), 1–20.

Dhawan, S. (2020). Online learning: A panacea in the time of COVID-19 crisis. *Journal of Educational Technology Systems, 49*(1), 5–22.

Finger, M., & Brand, S. B. (1999). The concept of the learning organization applied to the transformation of the public sector: Conceptual contributions for theory development. In: Easterby-Smith, M., Araujo, L. & Burgoyne, J., (Eds.). *Organizational Learning and the Learning Organization: Developments in Theory and Practice* (pp. 130–156): London: Sage.

Griffin, P., & Care, E. (2014). *Assessment and Teaching of 21st Century Skills: Methods and Approach.* Berlin: Springer.

Hancock, B., Okleford, E., & Windridge, K. (2007). *An introduction to qualitative research*: The NIHR RDS EM / YH.

Hoq, M. Z. (2020). E-Learning during the period of pandemic (COVID-19) in the kingdom of Saudi Arabia: An empirical study. *American Journal of Educational Research, 8*(7), 457–464.

Kalloo, R. C., Mitchell, B., & Kamalodeen, V. J. (2020). Responding to the COVID-19 pandemic in Trinidad and Tobago: Challenges and opportunities for teacher education. *Journal of Education for Teaching, 46*(4), 452–462.

Keeton, M. T. (2004). Best online instructional practices: Report of phase I of an ongoing study. *Journal of Asynchronous Learning Networks, 8*(2), 75–100.

Kim, K.-J., & Bonk, C. J. (2006). The future of online teaching and learning in higher education. *Educause quarterly, 29*(4), 22–30.

Lam, Y. J., Chan, C. M., Pan, H., & Wei, H. (2003). Differential developments of Taiwanese schools in organizational learning: Exploration of critical factors. *International Journal of Educational Management, 17,* 262–271.

Murphy, M. P. (2020a). Concluding Thoughts: What Can ('t) we Research About Emergency e-Learning? *PS: Political Science & Politics,* 54, 20–22.

Murphy, M. P. (2020b). COVID-19 and emergency eLearning: Consequences of the securitization of higher education for post-pandemic pedagogy. *Contemporary Security Policy,* 41, 1–14.

Ortega, M. C. G., & Falconer, L. (2015). Learning spaces in virtual worlds: Bringing our distance students home. *Journal of Applied Research in Higher Education, 7,* 83–98.

Qazi, A., Naseer, K., Qazi, J., AlSalman, H., Naseem, U., Yang, S., … Gumaei, A. (2020). Conventional to online education during COVID-19 pandemic: Do develop and underdeveloped nations cope alike. *Children and Youth Services Review, 119,* 105582.

Scull, J., Phillips, M., Sharma, U., & Garnier, K. (2020). Innovations in teacher education at the time of COVID19: an Australian perspective. *Journal of Education for Teaching, 46*(4), 497–506.

Stake, R. E. (2000). Case studies In: Denzin N. and Lincoln, Y. (Eds.). *Handbook of Qualitative Research* (pp. 435–455): Thousand Oaks, CA: Sage Publications.

Todorova, N., & Bjorn-Andersen, N. (2011). University learning in times of crisis: The role of IT. *Accounting Education, 20*(6), 597–599.

10 Electroencephalogram (EEG) Signals for Modern Educational Research

Fatima Bashir
Bahria University

Ahmed Ali
Sukkur IBA University

Toufique Akbar Soomro
Quaid-e-Awam University of Engineering, Science and Technology

M. Marouf and M. Bilal
Bahria University

Bhawani Shankar Chowdhry
Mehran University of Engineering & Technology

CONTENTS

DOI: 10.1201/9781003143796-10

10.1 BRAIN-COMPUTER INTERFACE (BCIs)

Over the past few years, BCIs have been shown to be the most promising of all neuroimaging techniques for modern educational Research. The BCIs assist and facilitate direct communication between computers and end-users. Typically, the BCI system consists of an input (neuronal activities), output (commands or states), and a controller that provides timing, offset, and on/off actions. It uses voluntary user neuronal activities to control the device by signals that interact with the surroundings; BCI does not require movement, nerves, or muscle interference to produce a command. This command can be given to games, digital painting, entertainment, artificial limbs, wheelchairs, etc. BCI-based devices have been successfully used in a number of fields, including clinical neurology, biomedical engineering, smart transport, marketing, education and manufacturing, etc. BCIs need brain signals to be processed in nonclinical environments, which poses a major challenge in managing these applications.

As shown in Table 10.1, several neuroimaging methods may be used with BCI, including Functional Near-Infrared Spectroscopy (fNIRS), Functional Magnetic Resonance Imaging (fMRI), Intracortical Neuron Recording, Electrocorticography (ECoG), Magnetoencephalography (MEG), and EEG. Comparatively, EEG has superior temporal resolution than other methods, and it is a comparatively new method in the field of consumer neuroscience. Due to its superiority over other methods (listed in Table 10.1), EEG signals are the most frequently used tools for educational studies. The merits of the EEG are discussed in detail in the following section.

10.2 ELECTROENCEPHALOGRAM (EEG)

EEG is the most common technique for the observation of neuronal activity in the form of electrical voltage variations by electrodes placed directly above the cortex. It is generally noninvasive; however, invasive electrodes can be employed as in electrocorticography. The electrodes are coupled together and arranged in a cap shape device. The EEG has the following main advantages compared to other techniques. Excellent temporal resolution (up to milliseconds), relatively simple to install and

TABLE 10.1
Comparison of BCI Neuroimaging Techniques

Neuroimaging Technique	Risk	Portability	Space Resolution	Temporal Resolution	Measurement Direct/Indirect	Measured Activity
NIRS	Noninvasive	Portable	~5 mm	~1 s	Indirect	Metabolic
FMRI	Non-invasive	Nonportable	~5 mm	~1 s	Indirect	Metabolic
Intracortical Neuron Recording	Invasive	Portable	~0.05 mm (LFP) ~0.1 mm (MUA) ~.5 mm (SUA)	~0.03 s	Direct	Electrical
ECoG	Invasive	Portable	~1 mm	~0.03 s	Direct	Electrical
MEG	No-invasive		~5mm	~0.05 s	Direct	Magnetic
EEG	Noninvasive	Portable	~10 mm	~0.05 s	Direct	Electrical

use, cheap, portable and noninvasive. Although spatial resolution and Signal to noise ratio (SNR) are not good compared to other techniques, EEG has the unique ability to provide continuous neuronal activity in real time. EEG is extremely useful for collecting clear and analytical evidence to help researchers better understand the underlying emotional and cognitive mechanisms in the production of choice-related information. In this chapter, therefore, the authors focused on studies based on EEG signals. Consider Figure 10.1 showing the 10-20 system, the international EEG electrode placement standard.

This method was established to adopt standardized recording to ensure that any results of the study (research or clinical) could be assembled, replicated, and compared and analyzed in a consistent manner across the scientific community. Typically, this method labels the sensors on the basis of their placement on the head. The space between the nearby sensors is 20%/10% of the head size. This method of electrode placement reduces the ratio of signal to noise and increases the reliability of the signal.

10.3 EEG SIGNAL COMPOSITION

The brain generates several neuronal activities that can be recorded as EEG signals. These signals can be composed of two kinds (I) field potentials and (II) rhythms. EEG signals can be further classified based on these two kinds of activities.

10.3.1 Field Potentials

The field potentials also known as transient activities mimic the action potentials of certain neurons in the form of spikes. These spikes can be identified by their operational properties, recurrence, shape, amplitude, frequency, and position. Two basic types of field potentials are event-related spectral perturbations (ERSPs) and event-related potentials (ERPs). ERPs have a very small magnitude. Therefore, EEG signals need to average over some duration to extract useful ERPs and remove error

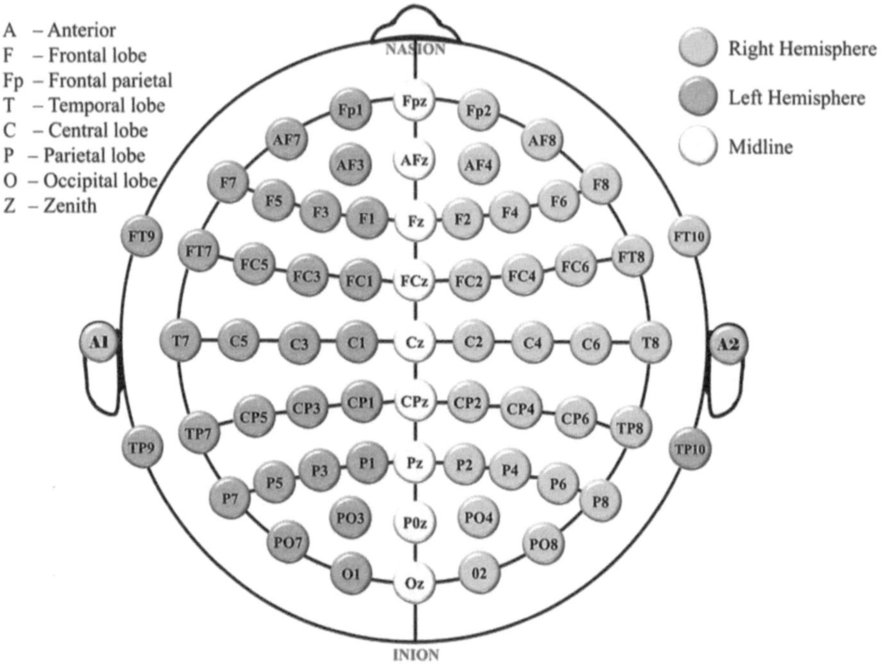

FIGURE 10.1 Electrode placement.

variations. Likewise, The ERSPs show the response to an event over a period of time and are comparable to the ERPs. However, ERPs divide neuronal activities into different waves to assess the dissimilarity that occurs over a period of time. The following section provides brief details of the standard ERPs used in educational studies.

10.3.1.1 Positive Slow Waves (PSW)

Positive slow waves (PSW) (Lin, Cross, Jones, & Childers, 2018) are associated with continued exposure to emotional visual stimulation and can be observed even after an emotional stimulus has occurred. Emotional factors and tasks, including arithmetic calculations, have also been shown to induce PSW. Although the magnitude of PSW may be an indication of the distribution of the source, variations in PSW shapes may suggest variations in the task-focused portion of the brain.

10.3.1.2 Feedback-Related Negativity (FRN)

FRN (Telpaz, Webb, & Levy, 2015) is the mid frontal negative potential associated with the decision of the subject. It occurs between 300 and 200 ms after the introduction of favorable and unfavorable products in response to passively observed products.

10.3.1.3 Frontal Negative 400 (FN400)

FN400 (Jones, Childers, & Jiang, 2012) is found in the familiarity-related front-center of the brain and occurs roughly 500–300 ms after a stimulus is started. FN400 has higher negative value for new words than for familiar and common words.

10.3.1.4 Negative 400

Negative 400 (N400) (Camarrone & Van Hulle, 2019) is a negative potential associated with quirkiness tests such as item titles.

10.3.1.5 Late Positive Potential

The late positive potential (Bastiaansen et al., 2018) indicates that either positive or negative valence requires exposure to emotional stimulation. Neutral, friendly, or negative perception stimuli may be translated as LPP. It occurs approximately 800-400 ms after the emergence of a stimulus.

10.3.1.6 P100/N100

Sensory-compulsory elements. Modulated by motivational salience, concentration, and focus relevance.

10.3.1.7 N200

N200 (Yadava et al., 2017) is an unfamiliarity-related negative potential that reaches its highest value between 200 and 350 ms after stimulation. Increased magnitude correlates with unfamiliar learning environment.

10.3.1.8 P300

Perhaps the most commonly examined ERP is P300 (Yadava et al., 2017) that has a positive magnitude value that occurs almost 300 ms just after the start of the cue due to internal decision-making. This indicates both the activity of working memory concentration and the transfer of reactions.

10.3.1.9 Contingent Negative Variation (CNV)

Contingent negative variance (CNV) is a surface negative slow potential that occurs when the second stimulus involves the motor response or decision of the participant in the time between the cortical reactions evoked by the stimulus pairs.

10.3.2 Brainwaves

Brainwaves, neural oscillations, or rhythms are an iterative pattern of neuronal activity. The EEG signal can be analyzed by dividing it into five frequency waves, as shown in Table 10.2, by these bands which are indicated as, γ, β α, θ, and δ waves. Table 10.2 also shows the stimulus usually associated with these bands, the brain area where these waves commonly occur, the shape, amplitude, and frequency ranges.

These frequency ranges have been associated with affective responses. The γ wave is mostly linked to the arousal effects, the β wave is linked to the variations during affective arousal.

The α wave in the frontal polar cortex distinguishes between negative and positive emotional valences.

The typical electrode positions used in the choice identification studies are summarized in Table 10.3. Touchette and Lee (2017) have shown that the α wave from positions F4 and F3 relate to the involuntary response of the customer to the items'

TABLE 10.2
Classification of the EEG Waves Based on Frequency

Wave	Frequency	Magnitude	Human Mental Stage
Gamma	(40–80) Hz	(3–5) μV or higher	Higher mental activity
Beta	(13–30) Hz	(2–20) μV or higher	Alertness, thinking
Alpha	(8–13) Hz	(30–50) μV or higher	Relaxed, not drowsy
Theta	(4–8) Hz	higher than 30 μV	Fantasy, dreaming
Delta	(0.1–4) Hz	(100–200) μV	Deep dreamless sleep

TABLE 10.3
Widely Used Locations of Sensors in Consumer Neuroscience Field for Choice Detection

Authors	Electrode Channel	Waves	Device System
Agarwal and Sivakumar (2019)	C4, Cz, Fz, Pz, P3, C3, P3 & F3	N200	Muse InteraXon
Touchette and Lee (2017)	F4 & Fp3	α	NeXus-10 biofeedback system Mind Media Inc
Vecchiato et al. (2011)	F1/F2, F7/F8, AF7/AF8 & Fp1/Fp2	α	BrainAmp, Brainproducts GmbH
Vecchiato et al. (2011)	F1/F2, F7/F8, AF7/AF8 & Fp1/Fp2	θ	BrainAmp, Brainproducts GmbH
Boksem and Smidts (2015)	CP5, C4, C1, C3, C5, FC4, FC2, FCz, FC1, FC3, F4, F2 & F1.	θ	64 active Ag–AgCl electrodes (Biosemi ActiveTwo
Boksem and Smidts (2015)	FCz, FC1, F2 & AFz,	β	64 active Ag–AgCl electrodes (Biosemi ActiveTwo

attractiveness. Vecchiato et al. (2011) have shown that asymmetric increments of the α and θ waves are associated with pleasant advertising in the left brain region at positions F1, F7, AF7, and FP1. Likewise, the same author (Vecchiato et al., 2011) also showed that asymmetric increments of the α and θ waves are linked to the viewing of displeasing advertising observed in the right brain region at positions F2, F8, AF8, and FP2. The spectral power of the α waves increases significantly for pleasant advertising at position F1 and for the unpleasant advertising at positions AF4 and AF8. In the contrary, for θ waves, enhanced neuronal activity develops at positions F3, AF8, and F2 for unpleasant advertising and at electrode Fp1 for pleasant advertising. Agarwal and Sivakumar (2019) related choice characteristics such as liking, emotions, and attention, to position C4, C3, Cz, Fz, Pz, P3, and F3. Boksem and Smidts (2015) have shown that the single choice is related to the activity of β in the (15–18) Hz span on the middle-frontal region at positions FCz, FC1, F2, and AFz. Besides, they demonstrated that the population choice is related to the (60–105) Hz θ span on the front-central region of electrodes CP5, C5, C4, C3, C1, FC4, FC2, FCz, FC4, FC1, F4, F2, and F1.

10.4 EEG IN EDUCATIONAL RESEARCH

Most EEG-related educational studies focus on experiments that replicate real-time circumstances; commercial-grade EEG headsets are deemed more satisfactory than clinical headsets. Commercial-grade EEG headsets are appreciated by teachers due to ease, wearability, single-channel use, and dry sensors. These advantages limit study subjects distress as well as anxiety while giving reliable and acceptable results. These systems provide easier solutions for tracking brain function when participants perform various learning activities. Brain-to-brain synchrony engagement time, attention & meditation, and other experiments are the four subgroups in which we separate various areas of study that have been done in actual classrooms or that have compared different forms of teaching/learning approaches.

10.4.1 Mind-to-Mind Synchrony

EEG-based hyperscanning methods were first used in social neuroscience experiments decades ago. In general, the studies seek to investigate the brain processes as participants perform various activities, including the synchronization/imitation of movement (Dumas, Nadel, Soussignan, Martinerie, & Garnero, 2010; Naeem, Prasad, Watson, & Kelso, 2012a, b; Tognoli, Lagarde, DeGuzman, & Kelso, 2007), music-related study (Babiloni et al., 2012, 2011; Lindenberger, Li, Gruber, & Müller, 2009), "Bridge-like" card game (Astolfi et al., 2010b; Astolfi et al., 2010; Babiloni et al., 2007, 2006), Chicken's Game (Astolfi et al., 2010b), and Prisoners Dilemma (Astolfi et al., 2009; Astolfi et al., 2011; Babiloni et al., 2007; Fallani et al., 2010). In order to reliably track variations in neuronal activity observed in various cortical regions, a huge number of electrodes and strict experimental protocols are required. The latest hyperscanning experiments, on the other hand, have shown that instructional research using consumer-grade EEG technologies to study brain synchronization in a real-world classroom setting has a bright future. Despite having lower electrode numbers, EEG devices of consumer grade show greater mobility and portability, which is considered more favorable in the real-world setting.

Mind-to-Mind synchrony is described as a correlation among the brainwaves of different people, under same learning environment. The common technique to calculate mind-to-mind synchrony is called total interdependence (Bevilacqua et al., 2019; Dikker et al., 2017; Rajagovindan & Ding, 2008; Wen, Mo, & Ding, 2012). During two entire semesters (spring and fall), the synchrony has been used to research social interactions, including contact between students and teachers in the real classroom (Bevilacqua et al., 2019). In a stereotypical classroom setting Bevilacqua et al. and Dikker et al. collected EEG signals from the students and their teachers and measured a synchrony (Bevilacqua et al., 2019; Dikker et al., 2017). Over the past few years, student-to-teacher, student-to student, and student-to-group learning activities have been assessed for brain-to-brain synchrony. The potential use of mind-to-mind synchrony as an indicator to research student interaction and social interactions was explored by Dikker et al. (2017). Mind-to-mind synchrony was shown to be connected to social dynamics, personality variations, and instructional approaches. Depending on the instructional approaches (lecture or video), student-to-group synchrony was higher in some groups. The students displayed a greater degree of student-to-student synchrony with their peers. The synchrony between the students and the instructor was primarily the subject of Bevilacqua et al. (2019). The EEG results, as well as student-teacher affection, self-reported commitment, and the students' quiz scores, were analyzed. According to the findings, the average synchrony of the group of students studying with videos was greater than that of the group studying with lectures. The synchrony of student-to-teacher connection was linked with their affection. In addition, the test scores of the students were also associated with affection, but not mind-to-mind synchrony. The results revealed how social dynamics, such as perceived affection, can predict cognitive effects, such as a student's academic success. Brain synchronization has drawn a lot of focus between students and between a student and a teacher and displays an interesting aspect that may be helpful in predicting how well students learn and the outcomes in the form of their academic success. Many protocols have been created using brain synchronization data, such as an effort to predict memory focus (Davidesco et al., 2019) and a combination of different consumer-grade EEG measuring sensors to track a wide group of participants, while they were listening to a speaker audibly reading a novel (Thammasan, Brouwer, Poel, & van Erp, 2020).

10.4.2 Engagement Time

Evaluating a person's engagement is significantly more difficult than evaluating the attention since it requires the general information on the person's engagement exercise. Treatment, satisfaction and a feeling of oddity or value may rely on the time spent in the work (Leiker et al., 2016; O'Brien & Toms, 2008). Thus, pupils' learning environment in the classroom or an online instructional exercise is additionally critical to the pupils accomplishment in learning. One estimation of engagement time that EMOVTIVE headset provide is an mnnm increase of the arousal and beta bands ascend alongside a decrease in alpha bands [34]. One other - EEG signals based methodology was developed in 2014 that assess students motivation (Ghergulescu & Muntean, 2014). For the assessment of the client's motivation relying upon the time, the Affective Suite's EmoEngine developed an extraordinary equation. The equations have outperform the standard survey procedure because this technique does not suffer from the interruption while acquiring data. This strategy is accordingly well suited to analyze the inspiration of the major part in the long arrangement of the game. It is all continuous and can give a clearer image of the move in the level of inspiration. The TimeonTask Threshold Computation (Ghergulescu & Muntean, 2014) technique was presented by Ghergulescu and Muntean in 2016. This new system is expected to help instructive game engineers track the investment of students during vivid exercises. The gadget assembles the measure of devotion, controlled by Affectiv Suite, and feeds the data in the data set alongside the time stamp of the occurrences. The system at that point decides naturally the vital qualities for moving specific occasions (when the estimation of an understudy's commitment diminishes to 5%, 10%, or 15%).

10.4.3 Attention and Meditation

The EEG approach has been appeared for quite a few years (Mulholland, 1968, 1969; Ray & Cole, 1985) to obtain and measure interest. From that point forward, researchers have recommended better approaches for recognizing and evaluating attention (Bajaj & Pachori, 2015; Bharti & Singh, 2015; Hamadicharef et al., 2009; Hu, Li, Sun, & Ratcliffe, 2016; Huang, Jung, & Makeig, 2007; Kelly, Dockree, Reilly, & Robertson, 2003; Srinivasan, Thorpe, Deng, Lappas, & D'Zmura, 2009). The methodologies differ with different extractions of attributes or calculations for de-nosing, for example, ghastly, principal component analysis (PCA), and changed wavelets. Momentum research is linked to the grouping of apparatus by a combination of space vector machine (SVM), Bayesian orders, concealed Markov models, Nearest neighbour models, and neural systems administration. Contingent upon the undertakings and the trial set-up, the arrangement precision is generally higher than 75%. The EEG channels provided by research grade gadgets range from 16 (BIOPAC Inc.) (Kelly et al., 2003) to 256 (BioSemi) (Huang et al., 2007). The precision of the information assembled and the attainable execution of EEG gadgets in market quality gets unsure with only one or a couple of organizations. Notwithstanding, an examination noticed that one of the EEG gadgets of the client grade can recognize subjects' advantage through the audit of crude EEG signals (Liu, Chiang, & Chu, 2013). To decide the center level, Liu et al. acquired crude EEG signals from 24 subjects with Attitude. The subjects needed to play out an English assignment in two separate conditions in which both the experimenters and the members were relied upon to show whether they were industrious. The energy estimation of every recurrence (alpha, beta, theta, and delta) alongside the proportion of alpha and beta conduct were obtained from five capacities. The proportion was consolidated based on a past examination in which the alpha action was shown to be loose and the ascent in the beta movement proposes more noteworthy attention (Hasegawa & Oguri, 2006). These qualities were utilized in the SVM characterization preparing. This lead to a grouping exactness of up to 76.82%. The equipment is viewed as precise to get authentic EEG flags despite the fact that the calculation does not work perfectly. To obtain and identify focus and reflection states, NeuroSky, a maker from Attitude and the MindBand, MindWave, and MindWave Smartphone offer clients a preassembled calculation. At the point when an individual chips away at visual capacities, for example, imaging portrayals of creatures, numbers, or specific items, the degree of fixation might be expanded. On the off chance that the client is diverted, the level may anyway be diminished. This calculation was, for the most part, used to compute the utility of different techniques for showing concentrate in instruction. When the client relaxes, the level of concentration increases, and if the client is distressed, the level of concentration decreases.

As we talked about in Section 10.2 previously, we feel that the genuine consideration of the buyer could be erroneous with such calculations. However, the developing/declining designs for following intentions are reliable (Maskeliunas, Damasevicius,

Martisius, & Vasiljevas, 2016) (Crowley, Sliney, Pitt, & Murphy, 2010; Patsis, Sahli, Verhelst, & De Troyer, 2013; Vourvopoulos & Liarokapis, 2014). Changes in the center degree are utilized by specialists as one of the estimation instruments. In a few training research reports, EEG gadgets have been utilized as approval strategies since 2014. These examination utilized the NeuroSky calculation to follow the fixation level of the members while doing the assignments identified with learning. Cell phone surveying comparisons, (Sun, 2014) the impact of versatile content showcases on buyer awareness, (Chen & Lin, 2016) the effect on people of various ages (grade school understudies and adults) (Ma & Wei, 2016; Wei & Ma, 2017), and the various types of book (ordinary, pictures, spring up, talking, and digital book), among others, were investigated. Portable democratic is a vivid path for a wide number of understudies to address the inquiry all the while on their cell phones. The degree of center when utilizing the cell phone surveys can be lower than the traditional clicker. In an analysis with text show, Chen and Lin show that there are no significant impacts on the clients' advantage for different types of text show, including static, intelligent, and blended (Chen & Lin, 2016). In a related survey, multi-tactile books or digital books are proposed to fit grade school understudies, though conventional books are liked for young ladies with advanced education. Despite the fact that EEG signals are an excellent indicator of concentration in training, different research groups have linked this property with other physiological effects. Chen and Wu (2015) utilized emWave to identify opinion and cognitive strain and students' yield in subjects to evaluate the proceeding with level of core interest. The emWave application gives a changeability calculation to cardiovascular cadence and pulse control. In paper-driven analyses to survey learning execution, the convention for the estimation of an intellectual charge depended on the psychological burden scale (Sweller, 1994). In addition to EEG-based studies determining the leveling of attention of people, Lai et al. used FaceReader, an outward appearance recognition programme (Huang, Liu, Lai, & Liu, 2017; Lai, Liu, Liu, & Huang, 2016). The objective of these investigations is, in any case, to check for changes in the EEG sign and outward appearance while members were seen to respond to questions effectively. The discoveries have demonstrated that understudies who got the correct reactions got more treatment than the individuals who got no prizes. No connection between' the helpful improvement and the learning yield has been found be that as it may. Albeit, most researchers just focus consideration, some additionally remember contemplation for their examination, as the Pride reflection and recognisable proof of NeuroSky is also outfitted (Lai et al., 2016; Shadiev, Huang, & Hwang, 2017; Shadiev, Wu, & Huang, 2017; Sun, 2014). Contemplation does subsequently not assume a significant part related to a center in an elaborate study hall and may just be joined with other execution evaluation factors. While the product given by NeuroSky is famous with scientists in the instructive area, numerous procedures to gauge consideration have been created. For example, a noticeable wavelet change (DWT) could separate a recurrence band (Shensa, 1992). Further usage of the hereditary calculation (Yang & Honavar, 1998) with the SVM classifier can be done for assortment of attributes (Chen, Wang, & Yu, 2017). Notwithstanding utilizing EEG buyer-grade instruments to test lessons/learning strategies, specialists have additionally made another stride, creating different systems to hold the premium of understudies and improve their learning execution. For the most part, these gadgets track the fixation level of the client and actuate an acoustic admonition when the client's consideration rises past a certain threshold (Chen & Wang, 2018; Kuo, Chu, & Tsai, 2017; Sun & Yeh, 2017). It can likewise make a connecting with specialist that can give appropriate criticism on the feeling and centralization of the client (Lin, Su, Chao, Hsieh, & Tsai, 2016). This intelligent interface has the limit not exclusively to focus on consideration (from brainwaves), yet in addition covers the passionate features of the clients utilizing Chinese semi-examination, EmguCV-based facial expression, and Q sensor skin yield, a wearable MIT feeling locator.

10.4.4 Other Studies

Aside from the standard EEG apparatuses mentioned earlier in this chapter, there are other research studies that focus on simultaneous EEG & eye tracking, for assessing intellectual burdens, the possibility that the student will fail the exam, and student frustration level during class etc. EEG was added to the neurolinguistic preliminaries to investigate distinctive neurophysiological cycles, including language treatment, movement of language, and language understanding (Elmer & Kühnis, 2016; Mollo, Pulvermüller, & Hauk, 2016; Tromp, Peeters, Meyer, & Hagoort, 2018). Be that as it may, the majority of the examinations have not yet considered the possibility of utilizing EEG client class instruments for etymological investigation. Notaro and Diamonds (2018) made an online German seminar on the electronic Duolingo interface for synchronous SEG, eye-following, conduction, and screen catch systems. EEG signals have been acquired at 250 Hz examining speeds by a 32-cycle OpenBCI (Cthon Biosensing) board. The standard 10-20 EEG gadget positions (Sharbrough, 1991) include eight situating cathodes. F3, Fz, F Cz, Cz, Pz, O1, and O2 were set. Gazepoint GP3 information at 60 Hz has been utilized to catch eye following information. Regularly enrolled are the offers of cursors, clunks, and screen catch. Students were instructed to complete four language lessons, which included "Basic 1," "Basic 2," "Use of The," and "Use of Phrases." The information has been proposed to give experiences into the formation of a site-based second language course from a neuroscience point of view. The subtleties can be found in Sinha et al.'s paper wherein they led an investigation utilizing Muse and EyeTribe to evaluate the perceptual pressing factor while appreciation, to perceive the challenges of clear literary material (Sinha, Roy, Chaki, De, & Saha, 2019). Two pieces of the trials were partitioned; worldwide and thorough evaluations. Just Muse was utilized in the worldwide audit to evaluate if the substance of the gathering was simple or hard to comprehend. It has been demonstrated that the entropy of the left transient zone was marginally more prominent when members read harder data, and when members partook in the less difficult substance, the entropy of the left frontal projection was higher. That demonstrated an impact on the language preparing of the left fleeting locale because of the multifaceted nature of text-based data. At the point when the more straightforward substance are examined, the eye development and its example have been considered and quickened, which influences the cerebrum actuation of the left front flap. Point by point research likewise explored the understanding conduct. New points of view utilizing EEG gadgets could help develop scholastic assessments across these individual practices and the confined cerebrum work. The student's ability to examine feelings using EEG will be influenced by the student's sense of outrage and energy. Following input from Intelligent Tutoring Systems (ITS), a PC framework prepared to give highlight support (Inventado, Legaspi, Suarez, & Numao, 2011), Inventado et al. attempted to anticipate understudy feelings, particularly dissatisfaction and excitement, by using the Emotiv EPOC (Ma & Wei, 2016).

The discoveries were resolved for the estimations of tumult and energy alongside ordinary varieties and mean, age and sex, five estimations of the Big Five character-assessing elements (John & Srivastava, 1999) and info styles (action change, arrangements assessment and clue). In the investigation, an assortment of calculations was likewise looked at: straight relapse, k nearest neighbor and support vector machine for expectation of enthusiastic levels. The findings show that direct relapse is associated with an increase in outrage, excitement, increases in dissatisfaction, and changes in excitation as opposed to other calculations with R2s of 0.462, 0.560, 0.627, and 0.484. The use of games for training has become popular among scientists and engineers, and it is regarded as appealing. In 2015, for example, LewiSpace was created as a 3D instructive game to display building graphs for various substance mixtures (Ghali, Ouellet, & Frasson, 2016).The game is planned to sort out whether the game framework is convoluted and to incorporate different clarifications or input for a more profound understanding. EEG signals from Emotiv EPOC (boredom, frustration, mediation and excitement, from Affectiv suite), student diameter from an eye tracker to calculate the participants workload, valence, and arousal, and facial expression from FaceReader software were among the features collected by the authors. The members have acquired robotized reports utilizing the Big five character poll. The qualities have been incorporated to gauge achievement or disappointment of each game mission. With strategic relapse, an AI calculation, the investigation accomplished 66% exactness. The Gaming related studies delivers deeper insights into the use of consumer-grade EEG devices in determining attention level. To the best of our knowledge, the most prevalent issue that has been identified demonstrates the ubiquity of attention research in the field. To improve the instructive climate, in any case, more intellectual measurements should be conducted. Emotive currently offers affirmation of the achievement pointers that comprise of pressure/disappointment, devotion, interest/valence, eagerness, fixation/consideration and unwinding/contemplation to help serve the clients. Both are fundamental contemplations and connected straightforwardly to tutoring. Such contemplations can be further explored, involving different field of study. Another test is that in genuine regular schools there are not many trials and no athletic exploration is distinguished. The greater part of the work was focused at web-based learning. We might want to check whether not simply PC/internet learning can supplant a teacher and additionally how it can help give satisfactory assets. More work is needed in ordinary homerooms or on those that assist the educator with giving an extra proper learning climate to improve learning proficiency and scholastic execution. Schooling research zeroed in on EEG is still at an early level, both in innovation and in data. These frameworks can offer a simple method to fuse, reduce the assembling time, limit expenses, and track other refined intellectual measurements soon. The measure of EEG-driven schooling concentrates in the client class would definitely increment fundamentally.

10.4.5 Education

The utilization of EEG has been adjusted for the field of instructive investigation to survey and track the viability of traditional and novel approaches or planning criticism-based structure to improve student's exhibition. An exertion on meta-dissecting the investigations on instruction utilizing compact EEG gadgets was tended to in 2017 (Xu & Zhong, 2018). Notwithstanding, more scientists have been exploring different avenues regarding such instruments as strategies for the investigation of instructive exploration since 2017, and this is getting more common. Rather than different features of training, the quantity of studies to survey and assess consideration is now strikingly high, like Xu and Zhong proposal's [235]. Clearly not less significant are other psychological segments, for example, feeling and mental depletion, which is not to be neglected. At present, EEG instruments of customer standard have been tried to ascertain and track such intellectual yields. In examination, most of studies set up web-based learning devices, solely measuring context reading abilities, and utilizing adults as test subjects rather than actual students in a traditional classroom. In the homeroom setting, a couple of examinations, including an investigation of mind synchrony, were performed for a full half year (Bevilacqua et al., 2019; Dikker et al., 2017). As of late, researchers have acquired consideration through internet learning concentrates because of their less complex methodology. In any event, it should not be dismissed when a teacher or supervisor is present. Studies recommend that contact among individuals can lead more to learning than collaboration with nonliving items, for instance a machine. A skilled and expert educator would be able to assess and evaluate the students' attention while adapting their teaching method and devoting extra attention to specific pupils who are struggling. BrainCo, Inc., a US-upheld start-up firm, publicizes one of its items as an EEG study hall follow-up headband. The business has worked together with a school in China, furnishing understudies and employees with its products. It very well may be valuable for educators and understudies to coordinate EEG gadgets and large information. The solutions, in particular, are still in the development stage, and the corporation has yet to fully introduce them into the market. One fundamental way might be the utilization of EEG during actual instruction for clients. However, current technology may not be capable of collecting a artifact-free and noise-free signals in a variety of situations without causing discomfort or restricting the individuals' movement.

10.5 EEG-BASED CLASSIFICATION TECHNIQUES FOR EDUCATION RESEARCH

10.5.1 Classification Methods

In the past few years, BCI systems have been designed by different EEG classification approaches through feature extraction and feature selection. These approaches include: neural network, linear classifiers, nearest neighbor classifiers, nonlinear Bayesian classifier, and classifier combinations. Two classifiers are included in linear classifier: linear discriminant analysis (LDA) and Support Vector Machine (SVM) that are popular for EEG based-BCIs (Lotte, Congedo, Lécuyer, Lamarche, & Arnaldi, 2007).

10.5.2 Linear Discriminant Analysis

LDA technique is more preferable for online BCI system because of low computational requirements and comprehensible nature. It uses a separate hyperplane to represent data in different classes (Duda & Hart, 2001; Fukunaga, 1990) with equal covariance matrix by assuming the normal distribution of data through LDA. This hyperplane helps to minimize the interclass variance and maximize the distance between means of two classes (Fukunaga, 1990). Possibly, this approach could be extended to multiple classes.

10.5.3 Support Vector Machine

Support vector machine (SVM) classifier helps distinguish the classes by using the hyperplane that separate one class from another by making boundaries. This classifier is mostly used in BCI applications because of overtraining, high dimensionality, and simplicity [2]. SVM is a supervised learning algorithm introduced in 1995 to solve problems related to binary classification of linear and or linear data. This classifier has solved a large number of BCI problems successfully (Blankertz, Curio, & Muller, 2002; Lotte et al., 2007; Rakotomamonjy, Guigue, Mallet, & Alvarado, 2005) but uses the 'kernel trick' to generate nonlinear decision boundaries that overcome the noise (Babiker, Faye, Mumtaz, Malik, & Sato, 2019). The previous literature emphasized that SVM performed well in terms of performance as compare to other classifiers (Lotte et al., 2007).

10.5.4 Artificial Neural Network

Neural networks (NN) are nonlinear classifiers that assemble a numbers of elements, known as neurons, and use the, to estimate nonlinear decision boundary. Linear classifiers with neural networks are effectively used in BCI research because of easy implementation and small training sets (Anderson & Sijercic, 1996; Hiraiwa, Shimohara, & Tokunaga, 1990). Multilayer perceptron (MLP) is a neural network that is extensively used in BCI and made up of numerous layers of neurons: some hidden layers and input and output layers (Bishop, 1995). The output of neurons of the previous layer is associated with each input layer neuron, while the input feature class is determined by the neurons of the output layer. Thus, MLP is one of the most popular neural network that is applied to solve most of the BCI problems such as multiclass or binary (Anderson & Sijercic, 1996; Palaniappan, 2005), asynchronous or synchronous (Chiappa & Bengio, 2003; Haselsteiner & Pfurtscheller, 2000) BCI.

10.5.5 Nonlinear Bayesian Classifiers

Nonlinear Bayesian classifiers are classifiers that assign feature vector by using the class through Bayes' rule and model the probability distribution of every class. In particular, these classifiers include Bayes quadratic classifiers and Hidden Markov model classifiers (HMMs). The Bayes quadratic contains the assumption of the standard distribution of data, which explains the name of the classifier by making the boundaries quadratic decision Although this classifier is not usually used for BCI, it has been useful for cognitive classification (Barreto, Frota, & de Medeiros, 2004; Keirn & Aunon, 1990).The HMM algorithms are ideally suitable for the time-series classification (Rabiner, 1989). BCI have specific time courses as it is driven by EEG components, and HMM has been used to identify temporal sequences of BCI features (Lotte et al., 2007; Obermaier, Guger, Neuper, & Pfurtscheller, 2001; Obermaier, Neuper, Guger, & Pfurtscheller, 2001) and also the raw EEG classification (Solhjoo, Nasrabadi, & Golpayegani, 2005). HMM within the BCI framework is not very widespread, but theses research has shown that they are appropriate classifiers for BCI technique.

10.5.6 K Nearest Neighbors

KNN is known as K nearest neighbors because of its ability to create classes by using the nearest neighbors based in feature vectors. These nearest neighbors are generally delineated by a metric distance (Blankertz et al., 2002). kNN can predict any function that allows nonlinear decision boundaries with the help of k training samples. Because of their responsive computational complexity (Friedman, 1997), which leads to the failure of many BCI experiments (Blankertz et al., 2002; Friedman, 1997; Schlögl, Lee, Bischof, & Pfurtscheller, 2005), the BCI community does not strongly support these algorithms. In addition, kNN may prove to be effective (Borisoff, Mason, Bashashati, & Birch, 2004) when feature vectors of low dimension are used in BCI systems.

10.5.7 Existing Research on the Machine-Learning Algorithm

The following section presents the different techniques used for classification and removing artifacts in the brain–computer interface, which has major research on both e-learning and classroom environments. Table 10.4 shows research on classification algorithms in different scenarios in BCI.

Lim and Chia (2015) presented the EEG signal methods that were processed in MATLAB environment. They used EEG with a single-electrode BCI to examine the efficiency of the cognitive pressure identification procedure. MATLAB was used for assessment, combined with DCT to decrease the data size to 2% of the source. Constants less than 2% DCT (discrete cosine transform) were used to train the ANN, KNN, and LDA pattern classification procedures. They determined that KNN with the use of 2% DCT has the highest classification accuracy of 72% as compared to other procedures.

The cognitive level of the students is a critical aspect to determine successful learning, as studied by (Zhou et al., 2017). Therefore, they aim to improve the attentiveness and engagement of students by creating the lessons proficiently. The instructors observed and adjusted the pace and subjects of the lecture with the cognitive level of students in a real-world classroom. They used EEG-based headset Emotiv Epoc+ to record the cognitive workload by designing and proposing a real-time passive BCI system. It is efficient and viable in an online digital environment, and the EEG signals were caught using two electrodes. They concluded that the existing model can monitor EEG signals and the cognitive loads can be classified using SVM while online course videos were being watched by the students.

TABLE 10.4
Research on Classification Algorithms in Different Scenarios in BCI

Application	Classifier	Environment	Feature	Accuracy (%)
Wang et al. (2013)	Naïve Bayes	MOOC Learning	Confusion Level	67
Lim and Chia (2015)	KNN	Classroom	Cognitive Stress	72
Chen, Wang and Yu (2017)	SVM	E-Learning	High and Low Attention Level	89.52
Zhou et al. (2017)	SVM	E-learning	Cognitive Workload and Attention	–
Myrden and Chau (2017)	SVM	E-Learning	Attention Level	84.8
Lin and Kao (2018)	SVM, Decision Tree, ANN	E-learning	Mental Effort (Understand, Not understand)	91.6, 95.6 and 93.1
Hu et al. (2016)	CFS+KNN	E-learning	Attention	80.84
Babiker et al. (2019), Kalpana, Ratnavelu, and Balasubramaniam (2019)	SVM,KNN	Classroom	Situational Interest Level	87.5 and 86.7

The viability of passive BCI using EEG to record the variations in the cognitive state on a particular experimental source was investigated (Myrden & Chau, 2017). To classify variations in attention state, frustration, and fatigue, the EEG signal qualities were redundantly classified using feature clustering procedures. The obtained accuracies in attention state, frustration, and fatigue were 84.8% ± 7.4%, 71.6% ± 5.6%, and 74.8% ± 9.1%, respectively. They concluded that for attention identification, posterior alpha-band activity is important, whereas posterior alpha band and frontal beta-band activity is important for frustration identification; the frontal and central electrodes are necessary for fatigue identification.

Further, Lin and Kao (2018) proposed a model based on cognitive load theory to monitor and label the mental state of an individual with viable EEG devices at the time of viewing online videos. Then they processed the EEG signals obtained from EEG device with different normalization patterns and time frames. Supervised learning classifiers, decision tree, ANN, and SVM were trained and tested, and obtained accuracies of 95.6%, 93.1%, and 91.6%, respectively. It shows that decision tree obtains the highest average accuracy. They determined that the proposed model was able to assist individual self-awareness of mental strengths efficiently in online learning environments.

Attention recognition through the processing of EEG while learning development was presented by Hu et al. (2016). They proposed a combined classification method consisting of data mining algorithms like CFS and KNN. They measured the performance 10 times with 3-fold cross-validation data. They assessed attention on three levels of high, neutral, and low with single valence using self-assessment design of self-report. The Correct classification rate (CCR) of CFS + KNN obtained was 80.84 percent 3.0 percent, and the performance rate was efficient.

The probability of perceiving situational attention by using EEG in the classroom was examined by (Babiker et al., 2019). The Ensemble Empirical Mode (EMD) was used to decompose the preprocessed EEG data. They used t-test and ROC to classify the IMFs based on their implications. The SVM and KNN classifiers were used by the particular features with ten cross-validations where SVM obtained higher accuracy of about 93.3% and 87.5% using features from the four EEG channels for two data sets while KNN obtained 87.5% and 86.7% using single EEG channel in the same datasets. They determined that the student's situational attention could be identified efficiently through gamma and delta bands.

Another study was conducted by Wang et al. (2013) to assess the confusion level of students while watching the course videos. A total of 20 educational videos were used in the experiment including 10 confusing and 10 nonconfusing ones, to classify student's confusion level. NeuroSky Mindset technique was used to collect EEG data while students were watching videos. After collecting data, nine students reported their level of confusion based on the user confusion level criteria. The Gaussian Nave Bayes classifier used different features such as attention, delta, mediation, theta, alpha, beta, and gamma to determine whether or not students become confused while watching videos. Predefine the level of confusion detected by the classifier to produce 0.67 accuracy.

10.6 CONCLUSION

The principles of electroencephalogram (EEG) signals in the field of education are discussed in this chapter of the book. Following that, we reviewed how EEG signals are currently used and consider how these signals might be used as tracking tools for educational scientists. Learning, success, promotion, motor skill development, e-learning, edutainment, immersive behavior, learning materials presentation behavior, and reading meaning were found to be the most common uses of EEG signals. We also discovered that EEG signals were mainly used to measure students' attention and reflection, that most EEG experiments lasted less than an hour, and that EEG sample sizes were minimal, and that university students were the largest research group. In the book chapter, we also addressed how existing EEG signal-based portable devices are yet to be optimized for classroom use. When used effectively, these instruments have a lot of potential as a source of quantitative findings, providing versatile, precise, objective, and fast results to supplement current classroom teaching testing techniques. In the final section of the book chapter, we explored the difficulties that EEG signal instruments pose, which involve major measurement errors and discomfort when used in large-scale and long-term studies. There are also certain research limitations: not just the lack of naturalistic classroom tests but also the lack of studies that delve at all areas of cognition. If the above-mentioned difficulties can be solved, EEG signals-based instruments may be commonly used in the field of education in the near future.

REFERENCES

Agarwal, M., & Sivakumar, R. (2019). *Cerebro: A Wearable Solution to Detect and Track User Preferences using Brainwaves.* Paper presented at the 5th ACM Workshop on Wearable Systems and Applications.

Anderson, C. W., & Sijercic, Z. (1996). *Classification of EEG signals from four subjects during five mental tasks.* Paper presented at the Solving Engineering Problems with Neural Networks: Proceedings of the Conference on Engineering Applications in Neural Networks (EANN'96).

Astolfi, L., Cincotti, F., Mattia, D., Fallani, F. D. V., Salinari, S., Marciani, M., ... He, B. (2009). *Estimation of the cortical activity from simultaneous multi-subject recordings during the prisoner's dilemma.* Paper presented at the 2009 Annual International Conference of the IEEE Engineering in Medicine and Biology Society.

Astolfi, L., Cincotti, F., Mattia, D., Fallani, F. D. V., Salinari, S., Vecchiato, G., ... Yuan, H. (2010a). *Simultaneous estimation of cortical activity during social interactions by using EEG hyperscannings.* Paper presented at the 2010 Annual International Conference of the IEEE Engineering in Medicine and Biology.

Astolfi, L., Toppi, J., Fallani, F. D. V., Vecchiato, G., Cincotti, F., Wilke, C. T., ... He, B. (2011). Imaging the social brain by simultaneous hyperscanning during subject interaction. *IEEE Intelligent Systems, 26*(5), 38.

Astolfi, L., Toppi, J., Fallani, F. D. V., Vecchiato, G., Salinari, S., Mattia, D., ... Babiloni, F. (2010b). Neuroelectrical hyperscanning measures simultaneous brain activity in humans. *Brain Topography, 23*(3), 243–256.

Babiker, A., Faye, I., Mumtaz, W., Malik, A. S., & Sato, H. (2019). EEG in classroom: EMD features to detect situational interest of students during learning. *Multimedia Tools and Applications, 78*(12), 16261–16281.

Babiloni, C., Buffo, P., Vecchio, F., Marzano, N., Del Percio, C., Spada, D., ... Perani, D. (2012). Brains "in concert": frontal oscillatory alpha rhythms and empathy in professional musicians. *NeuroImage, 60*(1), 105–116.

Babiloni, C., Vecchio, F., Infarinato, F., Buffo, P., Marzano, N., Spada, D., ... Perani, D. (2011). Simultaneous recording of electroencephalographic data in musicians playing in ensemble. *Cortex, 47*(9), 1082–1090.

Babiloni, F., Cincotti, F., Mattia, D., Fallani, F. D. V., Tocci, A., Bianchi, L., ... Astolfi, L. (2007). *High resolution EEG hyperscanning during a card game.* Paper presented at the 2007 29th Annual International Conference of the IEEE Engineering in Medicine and Biology Society.

Babiloni, F., Cincotti, F., Mattia, D., Mattiocco, M., Fallani, F. D. V., Tocci, A., ... Astolfi, L. (2006). *Hypermethods for EEG hyperscanning.* Paper presented at the 2006 International Conference of the IEEE Engineering in Medicine and Biology Society.

Bajaj, V., & Pachori, R. B. (2015). Detection of human emotions using features based on the multiwavelet transform of EEG signals. In: Hassanien A., & Azar A. (Eds.) *Brain-Computer Interfaces* (pp. 215–240): Cham: Springer.

Barreto, G. A., Frota, R. A., & de Medeiros, F. N. (2004). *On the classification of mental tasks: a performance comparison of neural and statistical approaches.* Paper presented at the Proceedings of the 2004 14th IEEE Signal Processing Society Workshop Machine Learning for Signal Processing, 2004.

Bastiaansen, M., Straatman, S., Driessen, E., Mitas, O., Stekelenburg, J., & Wang, L. (2018). My destination in your brain: A novel neuromarketing approach for evaluating the effectiveness of destination marketing. *Journal of destination marketing & management, 7*, 76–88.

Bevilacqua, D., Davidesco, I., Wan, L., Chaloner, K., Rowland, J., Ding, M., ... Dikker, S. (2019). Brain-to-brain synchrony and learning outcomes vary by student–teacher dynamics: Evidence from a real-world classroom electroencephalography study. *Journal of cognitive neuroscience, 31*(3), 401–411.

Bharti, S., & Singh, S. P. (2015). An enhanced feature extraction method and classification method of EEG signals using artificial intelligence. *International Journal of Computer Applications, 126*(10).

Bishop, C. M. (1995). *Neural Networks for Pattern Recognition*. Oxford: Oxford University Press.

Blankertz, B., Curio, G., & Muller, K.-R. (2002). Classifying single trial EEG: Towards brain computer interfacing. *Advances in Neural Information Processing Systems, 1*, 157–164.

Boksem, M. A. S., & Smidts, A. (2015). Brain responses to movie trailers predict individual preferences for movies and their population-wide commercial success. *Journal of Marketing Research, 52*(4), 482–492. doi:10.1509/jmr.13.0572.

Borisoff, J. F., Mason, S. G., Bashashati, A., & Birch, G. E. (2004). Brain-computer interface design for asynchronous control applications: improvements to the LF-ASD asynchronous brain switch. *IEEE Transactions on Biomedical Engineering, 51*(6), 985–992.

Camarrone, F., & Van Hulle, M. M. (2019). Measuring brand association strength with EEG: A single-trial N400 ERP study. *PLoS One, 14*(6), e0217125.

Chen, C.-M., & Lin, Y.-J. (2016). Effects of different text display types on reading comprehension, sustained attention and cognitive load in mobile reading contexts. *Interactive Learning Environments, 24*(3), 553–571.

Chen, C.-M., & Wang, J.-Y. (2018). Effects of online synchronous instruction with an attention monitoring and alarm mechanism on sustained attention and learning performance. *Interactive Learning Environments, 26*(4), 427–443.

Chen, C.-M., & Wu, C.-H. (2015). Effects of different video lecture types on sustained attention, emotion, cognitive load, and learning performance. *Computers & Education, 80*, 108–121.

Chen, C. M., Wang, J. Y., & Yu, C. M. (2017). Assessing the attention levels of students by using a novel attention aware system based on brainwave signals. *British Journal of Educational Technology, 48*(2), 348–369.

Chiappa, S., & Bengio, S. (2003). *HMM and IOHMM modeling of EEG rhythms for asynchronous BCI systems*. Retrieved from https://csilviavr.github.io/assets/publications/silvia04hmm.pdf

Crowley, K., Sliney, A., Pitt, I., & Murphy, D. (2010). *Evaluating a brain-computer interface to categorise human emotional response*. Paper presented at the 2010 10th IEEE International Conference on Advanced Learning Technologies.

Davidesco, I., Laurent, E., Valk, H., West, T., Dikker, S., Milne, C., & Poeppel, D. (2019). Brain-to-brain synchrony predicts long-term memory retention more accurately than individual brain measures. *bioRxiv*, 644047.

Dikker, S., Wan, L., Davidesco, I., Kaggen, L., Oostrik, M., McClintock, J., … Ding, M. (2017). Brain-to-brain synchrony tracks real-world dynamic group interactions in the classroom. *Current Biology, 27*(9), 1375–1380.

Duda, R. O., & Hart, P. E. (2001). *Stork. DG, Pattern Classification*. New York: John Willey and Sons.

Dumas, G., Nadel, J., Soussignan, R., Martinerie, J., & Garnero, L. (2010). Inter-brain synchronization during social interaction. *PLoS One, 5*(8), e12166.

Elmer, S., & Kühnis, J. (2016). Functional connectivity in the left dorsal stream facilitates simultaneous language translation: An EEG study. *Frontiers in human neuroscience, 10*, 60.

Fallani, F. D. V., Nicosia, V., Sinatra, R., Astolfi, L., Cincotti, F., Mattia, D., … He, B. (2010). Defecting or not defecting: how to "read" human behavior during cooperative games by EEG measurements. *PLoS One, 5*(12), e14187.

Friedman, J. H. (1997). On bias, variance, 0/1—loss, and the curse-of-dimensionality. *Data Mining and Knowledge Discovery, 1*(1), 55–77.

Fukunaga, K. (1990). *Statistical Pattern Recognition* (2nd ed.) (New York: Academic).

Ghali, R., Ouellet, S., & Frasson, C. (2016). LewiSpace: An exploratory study with a machine learning model in an educational game. *Journal of Education and Training Studies, 4*(1), 192–201.

Ghergulescu, I., & Muntean, C. H. (2014). A novel sensor-based methodology for learner's motivation analysis in game-based learning. *Interacting with Computers, 26*(4), 305–320.

Hamadicharef, B., Zhang, H., Guan, C., Wang, C., Phua, K. S., Tee, K. P., & Ang, K. K. (2009). *Learning EEG-based spectral-spatial patterns for attention level measurement.* Paper presented at the 2009 IEEE International Symposium on Circuits and Systems.

Hasegawa, C., & Oguri, K. (2006). *The effects of specific musical stimuli on driver~ s drowsiness.* Paper presented at the 2006 IEEE Intelligent Transportation Systems Conference.

Haselsteiner, E., & Pfurtscheller, G. (2000). Using time-dependent neural networks for EEG classification. *IEEE Transactions on Rehabilitation Engineering, 8*(4), 457–463.

Hiraiwa, A., Shimohara, K., & Tokunaga, Y. (1990). EEG topography recognition by neural networks. *IEEE Engineering in Medicine and Biology Magazine, 9*(3), 39–42.

Hu, B., Li, X., Sun, S., & Ratcliffe, M. (2016). Attention recognition in EEG-based affective learning research using CFS+ KNN algorithm. *IEEE/ACM Transactions on Computational Biology and Bioinformatics, 15*(1), 38–45.

Huang, R.-S., Jung, T.-P., & Makeig, S. (2007). *Multi-scale EEG brain dynamics during sustained attention tasks.* Paper presented at the 2007 IEEE International Conference on Acoustics, Speech and Signal Processing-ICASSP'07.

Huang, Y. M., Liu, M. C., Lai, C. H., & Liu, C. J. (2017). Using humorous images to lighten the learning experience through questioning in class. *British Journal of Educational Technology, 48*(3), 878–896.

Inventado, P. S., Legaspi, R., Suarez, M., & Numao, M. (2011). Predicting student emotions resulting from appraisal of its feedback. *Research & Practice in Technology Enhanced Learning, 6*(2), 107–133.

John, O. P., & Srivastava, S. (1999). *The Big-Five Trait Taxonomy: History, Measurement, and Theoretical Perspectives* (Vol. 2): Berkeley, CA: University of California Berkeley.

Jones, W. J., Childers, T. L., & Jiang, Y. (2012). The shopping brain: Math anxiety modulates brain responses to buying decisions. *Biological Psychology, 89*(1), 201–213.

Kalpana, M., Ratnavelu, K., & Balasubramaniam, P. (2019). An audio encryption based on synchronization of robust BAM FCNNs with time delays. *Multimedia Tools and Applications, 78*(5), 5969–5988.

Keirn, Z. A., & Aunon, J. I. (1990). A new mode of communication between man and his surroundings. *IEEE Transactions on Biomedical Engineering, 37*(12), 1209–1214.

Kelly, S., Dockree, P., Reilly, R., & Robertson, I. (2003). *EEG alpha power and coherence time courses in a sustained attention task.* Paper presented at the First International IEEE EMBS Conference on Neural Engineering, 2003. Conference Proceedings.

Kuo, Y.-C., Chu, H.-C., & Tsai, M.-C. (2017). Effects of an integrated physiological signal-based attention-promoting and English listening system on students' learning performance and behavioral patterns. *Computers in Human Behavior, 75*, 218–227.

Lai, C.-H., Liu, M.-C., Liu, C.-J., & Huang, Y.-M. (2016). Using positive visual stimuli to lighten the online learning experience through in class questioning. *International Review of Research in Open and Distributed Learning, 17*(1), 23–41.

Leiker, A. M., Miller, M., Brewer, L., Nelson, M., Siow, M., & Lohse, K. (2016). The relationship between engagement and neurophysiological measures of attention in motion-controlled video games: A randomized controlled trial. *JMIR Serious Games, 4*(1), e4.

Lim, C.-K. A., & Chia, W. C. (2015). Analysis of single-electrode EEG rhythms using MATLAB to elicit correlation with cognitive stress. *International Journal of Computer Theory and Engineering, 7*(2), 149.

Lin, F.-R., & Kao, C.-M. (2018). Mental effort detection using EEG data in E-learning contexts. *Computers & Education, 122*, 63–79.

Lin, H.-C. K., Su, S.-H., Chao, C.-J., Hsieh, C.-Y., & Tsai, S.-C. (2016). Construction of multi-mode affective learning system: Taking affective design as an example. *Journal of Educational Technology & Society, 19*(2), 132–147.

Lin, M.-H. J., Cross, S. N., Jones, W. J., & Childers, T. L. (2018). Applying EEG in consumer neuroscience. *European Journal of Marketing, 52*, 66–91.

Lindenberger, U., Li, S.-C., Gruber, W., & Müller, V. (2009). Brains swinging in concert: cortical phase synchronization while playing guitar. *BMC Neuroscience, 10*(1), 1–12.

Liu, N.-H., Chiang, C.-Y., & Chu, H.-C. (2013). Recognizing the degree of human attention using EEG signals from mobile sensors. *Sensors (Basel), 13*(8), 10273–10286.

Lotte, F., Congedo, M., Lécuyer, A., Lamarche, F., & Arnaldi, B. (2007). A review of classification algorithms for EEG-based brain–computer interfaces. *Journal of Neural Engineering, 4*(2), R1.

Ma, M.-Y., & Wei, C.-C. (2016). A comparative study of children's concentration performance on picture books: Age, gender, and media forms. *Interactive Learning Environments, 24*(8), 1922–1937.

Maskeliunas, R., Damasevicius, R., Martisius, I., & Vasiljevas, M. (2016). Consumer-grade EEG devices: Are they usable for control tasks? *PeerJ, 4*, e1746.

Mollo, G., Pulvermüller, F., & Hauk, O. (2016). Movement priming of EEG/MEG brain responses for action-words characterizes the link between language and action. *Cortex, 74*, 262–276.

Mulholland, T. (1968). The concept of attention and the EEG alpha-rhythm. *Electroencephalography and Clinical Neurophysiology, 24*(2), 188.

Mulholland, T. (1969). The concept of attention and the electroencephalographic alpha rhythm. In: Evans C. R. & Mulholland T. B. (Eds.) *Attention in Neurophysiology* (pp. 100–127): London: Butterworths.

Myrden, A., & Chau, T. (2017). A passive EEG-BCI for single-trial detection of changes in mental state. *IEEE Transactions on Neural Systems and Rehabilitation Engineering, 25*(4), 345–356.

Naeem, M., Prasad, G., Watson, D. R., & Kelso, J. S. (2012a). Electrophysiological signatures of intentional social coordination in the 10–12 Hz range. *NeuroImage, 59*(2), 1795–1803.

Naeem, M., Prasad, G., Watson, D. R., & Kelso, J. S. (2012b). Functional dissociation of brain rhythms in social coordination. *Clinical Neurophysiology, 123*(9), 1789–1797.

Notaro, G. M., & Diamond, S. G. (2018). Simultaneous EEG, eye-tracking, behavioral, and screen-capture data during online German language learning. *Data in brief, 21*, 1937–1943.

O'Brien, H. L., & Toms, E. G. (2008). What is user engagement? A conceptual framework for defining user engagement with technology. *Journal of the American society for Information Science and Technology, 59*(6), 938–955.

Obermaier, B., Guger, C., Neuper, C., & Pfurtscheller, G. (2001). Hidden Markov models for online classification of single trial EEG data. *Pattern Recognition Letters, 22*(12), 1299–1309.

Obermaier, B., Neuper, C., Guger, C., & Pfurtscheller, G. (2001). Information transfer rate in a five-classes brain-computer interface. *IEEE Transactions on Neural Systems and Rehabilitation Engineering, 9*(3), 283–288.

Palaniappan, R. (2005). *Brain computer interface design using band powers extracted during mental tasks*. Paper presented at the Conference Proceedings. 2nd International IEEE EMBS Conference on Neural Engineering, 2005.

Patsis, G., Sahli, H., Verhelst, W., & De Troyer, O. (2013). *Evaluation of attention levels in a tetris game using a brain computer interface*. Paper presented at the International conference on user modeling, adaptation, and personalization.

Rabiner, L. R. (1989). A tutorial on hidden Markov models and selected applications in speech recognition. *Proceedings of the IEEE, 77*(2), 257–286.

Rajagovindan, R., & Ding, M. (2008). Decomposing neural synchrony: Toward an explanation for near-zero phase-lag in cortical oscillatory networks. *PLoS One, 3*(11), e3649.

Rakotomamonjy, A., Guigue, V., Mallet, G., & Alvarado, V. (2005). *Ensemble of SVMs for improving brain computer interface P300 speller performances*. Paper presented at the International conference on artificial neural networks.

Ray, W. J., & Cole, H. W. (1985). EEG alpha activity reflects attentional demands, and beta activity reflects emotional and cognitive processes. *Science, 228*(4700), 750–752.

Schlögl, A., Lee, F., Bischof, H., & Pfurtscheller, G. (2005). Characterization of four-class motor imagery EEG data for the BCI-competition 2005. *Journal of Neural Engineering, 2*(4), L14.

Shadiev, R., Huang, Y.-M., & Hwang, J.-P. (2017). Investigating the effectiveness of speech-to-text recognition applications on learning performance, attention, and meditation. *Educational Technology Research and Development, 65*(5), 1239–1261.

Shadiev, R., Wu, T.-T., & Huang, Y.-M. (2017). Enhancing learning performance, attention, and meditation using a speech-to-text recognition application: Evidence from multiple data sources. *Interactive Learning Environments, 25*(2), 249–261.

Sharbrough, F. (1991). American Electroencephalographic Society guidelines for standard electrode position nomenclature. *Journal of Clinical Neurophysiology, 8*, 200–202.

Shensa, M. J. (1992). The discrete wavelet transform: wedding the a trous and Mallat algorithms. *IEEE Transactions on Signal Processing, 40*(10), 2464–2482.

Sinha, A., Roy, D., Chaki, R., De, B. K., & Saha, S. K. (2019). Readability analysis based on cognitive assessment using physiological sensing. *IEEE Sensors Journal, 19*(18), 8127–8135.

Solhjoo, S., Nasrabadi, A. M., & Golpayegani, M. R. H. (2005). *Classification of chaotic signals using HMM classifiers: EEG-based mental task classification*. Paper presented at the 2005 13th European Signal Processing Conference.

Srinivasan, R., Thorpe, S., Deng, S., Lappas, T., & D'Zmura, M. (2009). *Decoding attentional orientation from EEG spectra*. Paper presented at the International Conference on Human-Computer Interaction.

Sun, J. C.-Y. (2014). Influence of polling technologies on student engagement: An analysis of student motivation, academic performance, and brainwave data. *Computers & Education, 72*, 80–89.

Sun, J. C.-Y., & Yeh, K. P.-C. (2017). The effects of attention monitoring with EEG biofeedback on university students' attention and self-efficacy: The case of anti-phishing instructional materials. *Computers & Education, 106*, 73–82.

Sweller, J. (1994). Cognitive load theory, learning difficulty, and instructional design. *Learning and Instruction, 4*(4), 295–312.

Telpaz, A., Webb, R., & Levy, D. J. (2015). Using EEG to predict consumers' future choices. *Journal of Marketing Research, 52*(4), 511–529.

Thammasan, N., Brouwer, A.-M., Poel, M., & van Erp, J. (2020). Interpersonal EEG synchrony while listening to a story recorded using consumer-grade EEG devices. In: Davis F., Riedl R., vom Brocke J., Léger P.M., Randolph A., Fischer T. (Eds.). *Information Systems and Neuroscience* (pp. 253–259): Cham: Springer.

Tognoli, E., Lagarde, J., DeGuzman, G. C., & Kelso, J. S. (2007). The phi complex as a neuromarker of human social coordination. *Proceedings of the National Academy of Sciences, 104*(19), 8190–8195.

Touchette, B., & Lee, S.-E. (2017). Measuring neural responses to apparel product attractiveness: An application of frontal asymmetry theory. *Clothing and Textiles Research Journal, 35*(1), 3–15.

Tromp, J., Peeters, D., Meyer, A. S., & Hagoort, P. (2018). The combined use of virtual reality and EEG to study language processing in naturalistic environments. *Behavior Research Methods, 50*(2), 862–869.

Vecchiato, G., Toppi, J., Astolfi, L., De Vico Fallani, F., Cincotti, F., Mattia, D., … Babiloni, F. (2011). Spectral EEG frontal asymmetries correlate with the experienced pleasantness of TV commercial advertisements. *Medical & Biological Engineering & Computing, 49*(5), 579–583. doi:10.1007/s11517-011-0747-x.

Vourvopoulos, A., & Liarokapis, F. (2014). Evaluation of commercial brain–computer interfaces in real and virtual world environment: A pilot study. *Computers & Electrical Engineering, 40*(2), 714–729.

Wang, H., Li, Y., Hu, X., Yang, Y., Meng, Z., & Chang, K.-M. (2013). *Using EEG to Improve Massive Open Online Courses Feedback Interaction*. Paper presented at the AIED Workshops.

Wei, C.-C., & Ma, M.-Y. (2017). Influences of visual attention and reading time on children and adults. *Reading & Writing Quarterly, 33*(2), 97–108.

Wen, X., Mo, J., & Ding, M. (2012). Exploring resting-state functional connectivity with total interdependence. *NeuroImage, 60*(2), 1587–1595.

Xu, J., & Zhong, B. (2018). Review on portable EEG technology in educational research. *Computers in Human Behavior, 81*, 340–349.

Yadava, M., Kumar, P., Saini, R., Roy, P. P., & Dogra, D. P. (2017). Analysis of EEG signals and its application to neuromarketing. *Multimedia Tools and Applications, 76*(18), 19087–19111.

Yang, J., & Honavar, V. (1998). Feature subset selection using a genetic algorithm. In: Liu H., Motoda H. (eds.) *Feature Extraction, Construction and Selection* (pp. 117–136): Boston, MA: Springer.

Zhou, Y., Xu, T., Cai, Y., Wu, X., & Dong, B. (2017). *Monitoring cognitive workload in online videos learning through an EEG-based brain-computer interface*. Paper presented at the International Conference on Learning and Collaboration Technologies.

11 Impact of Students' Psychological Distress on Their Academic Performance Through E-Learning Readiness

Gul Afshan
Sukkur IBA University

Maryam Kalhoro
University of Sindh

Nasima Zainulabidin
Sukkur IBA University

CONTENTS

DOI: 10.1201/9781003143796-11

11.1 INTRODUCTION

COVID-19, caused by severe acute respiratory coronavirus-2 disease (SARS-CoV-2), was first identified in the city of Wuhan, China, in December 2019 (Temsah et al., 2020). After identification of COVID-19, it spread around the world. This spread of the fearful disease made the World Health Organization (WHO) declare the situation a global pandemic on 11 March 2020 (Kapasia et al., 2020). Universities have been severely affected globally due to this pandemic of COVID-19. To avoid this discontinuity of education during this pandemic, universities tried to replace traditional learning mechanism with online learning system within a very short span of time. According to the survey of IAU (2020), more than 1.5 billion students have suffered because of the closure of schools and universities in COVID-19 pandemic. The novel coronavirus has directly impacted the performance of students and institutions (Aucejo et al., 2020). Consequently, this online leaning approach has left the teachers and students facing numerous challenges, and they are facing psychological problems (Alam, 2020; Bao, 2020). Still, many countries have not completely accepted this e-learning mechanism (Thongsri, Shen and Bao, 2019). The study of Jaeger and Blaabaek (2020) revealed that those students who are from poor families cannot afford to avail such facilities of online learning, thus resulting in discrimination. Beaunoyer et al. (2020) have also found the digital disparity among students in this period of COVID-19 that has been the cause of stress.

Although, most of the educational institutions have adopted e-learning system for maintaining their reputation (Yen and Mohammed, 2020; Zhou et al., 2020), its benefits to the students of low-income families and those in rural and remote areas in developing countries are not clear. Simultaneously, as stated by Adam et al. (2020), the disproportionate internet usage and its cost hinder access to these online classes. Thus, directly and indirectly, COVID-19 has caused psychological distress among students. This has negatively affected students' academic performance. However, it is important to determine the underlying factor in the relationship between psychological distress and academic performance. We believe that e-learning readiness may play the role of mediating mechanism in explaining this relationship. Studies have found very low success rate of e-learning (Yilmaz, 2017; Wei and Chou, 2020).

The COVID-19 pandemic enforces the institutions to reduce social interactions. In such a case, academic leader's social assistance and transformation of e-learning was the need of the time. In times of such a crisis such as the pandemic, leadership has to play its part (Fernandez and Shaw, 2020; Paixão et al., 2020; Dirani et al., 2020). Not only were higher educational regulatory bodies confused, but every stakeholder, including students, were in a state of confusion to continue their learning process in such a critical time of health crisis. Despite that, most of the HEIs in Pakistan, particularly private, were able to manage their teaching-learning process. However, there is a lack of literature to determine whether leadership was successful in delivering the appropriate support and technological resources to faculty and students to continue this process of teaching and learning. Therefore, there is a need to examine the role of change-oriented leadership in HEIs in response to COVID-19 to reduce psychological distress among students to facilitate their e-learning readiness. It is important because psychological distress and mental health disorders are commonly observed among the university students who are using online learning platform with low level of social support (Farrer et al., 2016; Fawzy and Hamed, 2017; Cao et al., 2020).

This study aims to contribute to the recent literature in many ways. First, although researchers have studied COVID-19 and its impacts on education, there is a lack of research on individual factors affecting e-learning and the role of leadership in HEIs in crisis periods. Thus, this study's main theme is to examine the effect of psychological distress caused by COVID-19 to influence students' readiness for e-learning and academic performance. Second, most of the research on COVID-19 effects on HEIs have been conducted in the western part of the world (Greenhow et al., 2020; Pather et al., 2020; Sá and Serpa, 2020). There is need to study how psychological distress has effected academic performance of students in the developing context. Thus, this study aims to fills this gap to investigate the hypothesized relationship in the Pakistani academic context. Third, the leadership's role in facilitating students' e-learning readiness and academic performance may provide an understanding of whether change-oriented leadership facilitates HEIs to sustain their educational systems. This empirical research work takes the perspective of Pakistani HEIs and may add significant value to the ever-growing domain of e-learning pedagogy. Moreover, besides theoretical contributions, this research has practical implications for academicians, policymakers, and managers, especially for HEIs in developing countries.

11.2 LITERATURE REVIEW AND HYPOTHESIS DEVELOPMENT

The educational and health system of Pakistan during the pandemic is not different from that in the rest of the developing nations. The global security index has ranked Pakistan among the countries with poor preparation for pandemics. Pakistan's health system lacks basic infrastructure, public awareness and communication about emergencies, infection control, and the health instruments to cope with many severe diseases. People in Pakistan, including students, are experiencing anxiety, stress, and depression due to pandemics (Riaz et al., 2021; Rehman et al., 2020). Importantly, illiterate people, compared to literate ones, are reluctant to accept that COVID-19 is a real and dangerous thing (Gul, 2021). This shows that awareness and knowledge are important indicators for change and improvement in behavior (Nadeem and Khaliq, 2021), particularly in HEIs (Sahibzada et al., 2020a, b)

In response to COVID-19, educational schools closed in mid-March, 2020, which transformed the need to shift to online learning. However, it was the best decision only for students well equipped with all the technological resources (Farooq et al., 2020; Rehiem, 2021; Joshi et al., 2020). In developing countries like Pakistan, most of the students do not even have a smartphone or computers for online classes. It also specifies that besides lack of resources, faculty and staff are also not adept at using online teaching technology. Moreover, the government schools mostly have students from low-income families, who already lack such resources, which disturbs them and hinders online education (Roy et al., 2020; Alderman et al., 2001).

Purchasing gadgets in a developing country like Pakistan is one of the major financial issues. Due to a collectivistic society, parents or elders are the sole people responsible for their children's resources until they complete student life. Remote areas lack internet access or have slow-speed network along with heavy load shedding of electricity. All these issues during COVID-19 have contributed to students' psychological issues and have negatively affected their readiness and participation in online classes. Although the Government of Pakistan, in the latest educational policy, has emphasized the usage of technology and e-learning in schools and universities, the teachers have also been suggested to integrate information and technological tools in classrooms. However, its integration effectiveness depends on resource availability and skill competencies (Asad et al., 2020).

Considering the above discussion and the context of Pakistan, the hypothesized relationships are explained below.

11.2.1 Psychological Distress and Academic Performance

Research has shown that epidemics and outbreaks in society have profound impacts on physical health and mental health (Khan et al., 2020; Mamun et al., 2021; Ahmad et al., 2020). Many countries in the world adopted the measures to tackle the physical and psychological effects of the pandemic. Due to lack of mobility, people were pushed toward isolation, depression, and mental illness, which results in emotional and behavioral changes. The most vulnerable group was students, due to academic uncertainty, hampered education system, and closure of each activity. However, much less attention was paid to their mental health during the COVID-19 pandemic because they were adopting a new learning method (Wang et al., 2020). Consequently, this COVID-19 has created a psycho-emotional chaotic situation in many countries and has led to a sharp rise of mental health problems in learners, such as anxiety, depression, stress, sleep disorder, and fear (Akinkugbe et al., 2020; Yassin et al., 2021; Hassan and Bao, 2020).

Problems related to mental health disturb people, but it is found that university students are on the higher susceptibility side for psychological disorders (Wang et al., 2020; Odriozola-González, 2020; Aristovnik et al., 2020). This is because university students encountered various tasks simultaneously like academic assignments, deadlines, presentation, workload, and anxiety to maintain their prestige and performance, faculty and parental expectations, etc. Recent literature by Jaeger and Blaabaek (2020) discloses that most students have uneven learning opportunities and chances, which results in discrimination in families. Beaunoyer et al. (2020) investigated the digital disparities during the COVID-19 pandemic, where most of the academic establishments are adopting the online platform, and found that it is difficult for low-income families and especially those who are staying in remote areas to access online classes as well as to get benefits with this platform. Similarly, Baticulon et al. (2020) identified that there are mainly five types of hurdles to e-learning such as technological, personal, family, institutional, and communities, and only 41% of students thought they could physically and mentally participate in online study. The said empirical study identified the impact of COVID-19 on college students as they suffer from psychological distress due to fear of academic loss, internet issues, lack of resources, etc.

Noor et al. (2020), in their qualitative study on a Pakistani student sample, found that improper study environment, inequality in the availability of gadgets, insufficient online study material, no access or slow internet speed, and load shedding are among the major issues that students face. They also found higher sleep deprivation, weak eyesight, and behavioral issues among students in lockdown areas. These kinds of mental health issues due to COVID-19 have affected students' online learning and performance in Pakistan (Shakeel et al., 2020).

Research in other developing countries has also found such issues among the students. For instance, Majumdar et al. (2020) found that sleep disruption, depression, somatic pain, and increased screen exposure were the major concerns among students in India. The mental health and even suicide cases during COVID-19 have been on the rise in developing countries (Agarwal et al., 2021; Lathabhavan, 2020).

Considering the above discussion, it can be said that psychological distress during COVID-19 may negatively influence their academic performance. Thus, it is hypothesized that

H1: Student's psychological distress during COVID-19 is negatively associated with their academic performance.

11.2.2 Mediation of Student's e-Learning Readiness

Readiness is a very important aspect of the teaching-learning process, and it plays a noteworthy role in learning (Hung et al., 2010; Yilmaz, 2017). Mainly, modification in behavior takes place based on the readiness of the learner. e-learning readiness, as defined by Kaur and Abbas (2004), is an individual's ability to utilize e-resources and technological tools to enhance the learning quality. Literature has conceptualized e-learning readiness with self-efficacy related to the computer, internet, online communication, and motivation to learn electronically (Demir, 2015, Hao, 2016).

Due to COVID-19, a teaching-learning process is conducted online, entirely different from formal classroom settings and physical learning environments. However, the readiness to educate for the educator and learn for the learner should not be overlooked in this different medium. e-learning helps students, teachers, and other professionals enhance their information technology skills and is essential in the workplace. However, students are facing digital inequality (Li and Lalani, 2020; Tam and El-Azar, 2020) and are afraid, angry, and helpless (Butz et al., 2015) due to inadequate resources for online learning during COVID-19 (Qazi et al., 2020; Noor et al., 2020). These and similar psychological distress levels have weakened students' ability to be ready for online learning with self-efficacy and motivation.

Recent research has shown lack of physical interaction and data security issues to hinder e-learning readiness among students (Phutella and Dwivedi, 2020). No doubt, the pandemic has created an opportunity to update and upgrade the technological skills among students in Pakistan; however, internet accessibility issues have made this pandemic stress even worse (Khan and Abid, 2021). All in all, for successful e-learning completion, students' e-learning readiness is highly important (Demir, 2015; Wei and Chou, 2020).

Based on the above studies, it is expected that due to psychological distress among students for several reasons during COVID-19, students would not be ready for e-learning, which in turn affects their academic performance. Thus, we hypothesize the following relationship:

H2: Readiness for e-learning mediates the relationship between student's psychological distress and academic performance.

11.2.3 The Moderating Role of Change-Oriented Leadership

Leaders in organizations play an important part in employee's performance (Afshan and Serrano-Archmi, 2020; Raaper and Brown, 2020; Adebola et al., 2020; Eveland, 2020), particularly in HEIs (Sahibzada et al., 2020a, b). A leader's role becomes more significant in crisis times (Fernandez and Shaw, 2020; Dirani et al., 2020). In the era of COVID-19, when the nations declared lockdown completely, it was one of the hardest challenges for some leaders to resume the process of education. The studies shows that the lack of competencies and limited resources hinder management from motivating teachers to implement information and computer technology in teaching-learning (Asad et al., 2020). It is seen during this pandemic that disasters can make or break a leader. However, it has been observed that educational leaders successfully disseminated the information, strategies, and approaches to revive the education system by exploring new technology avenues. In such times of crisis of COVID-19, change-oriented leadership is the need of the time. Change-oriented leaders are constantly motivated to work for innovation, change, and creativity (Kotter, 2012). These characteristics make a leader a promoter of organizational development. Under such supervision, employees are better able to determine innovative practices to tackle the change (Kotter, 2012). Change-oriented leaders work in the best interest of the organization (Bass and Bass, 2008).

Especially at this time of COVID-19, decision-making is the most critical aspect for overcoming any organization's treacherous moments. In order to tackle the challenges of COVID-19, it is the foremost duty of academic leaders to support academia to provide training and resources to faculty and staff, such as a toolbox of intellectual stimulation, idealized influence, and inspiration (Fernandez and Shaw, 2020). The prime motive at this time is to come together for the well-being of learners (Dever and Justice, 2020).

Keeping in view the critical role of leaders in times of crisis such as COVID-19 and the related literature, it is expected that change-oriented leadership may moderate the relationship between psychological distress and e-learning readiness. Moreover, it is also expected that the indirect effect of student's psychological distress on academic performance via e-learning readiness may depend on the value of change-oriented leadership, thus making the research framework a moderated mediation model (Figure 11.1).

Thus, we hypothesize the following relationship:

H3: Change-oriented leadership moderates the relationship between psychological distress and e-learning readiness.

H4: Change-oriented leadership moderates the mediating effect of e-learning readiness in the relationship between psychological distress and academic performance such that at high value of change-oriented leadership, the indirect effect would be weaker.

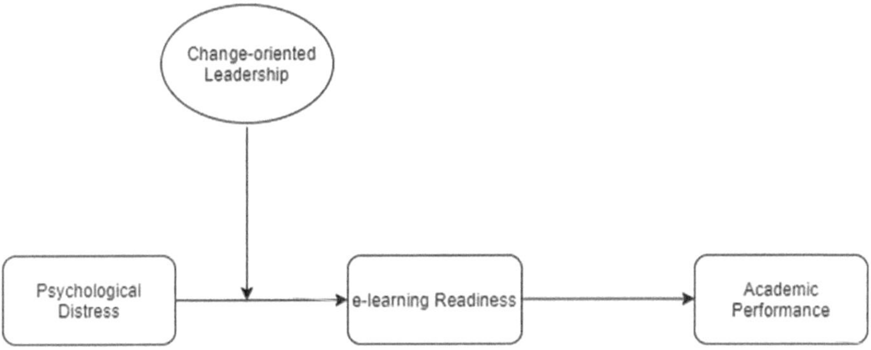

FIGURE 11.1 Research model depicting relationship among the study constructs.

11.3 METHODOLOGY

11.3.1 Procedure

The students from HEIs in Pakistan were chosen to collect data about their psychological distress, readiness for e-learning, and academic performance during COVID-19. A total of 264 students responded to the questionnaire. The sample consisted of 23% female students ($n = 60$), and 77% male students ($n = 204$). Students belonged to bachelor's and master's level courses. Table 11.1 shows the descriptive statistics of the sample.

11.3.2 Measuring Instruments

The data was collected through an online survey questionnaire due to the pandemic. The survey consists of well-established scales adopted from the literature. It included the psychological distress, students' readiness for e-learning, academic performance, and change-oriented leadership scales. All the scales were measured using a 1–5 Likert scale ranging from strongly disagree to strongly agree. The scales were slightly modified based on the objective of the study.

Psychological distress: The 5-item short version of psychological distress scale developed by Hassan and Bao (2020) was used in this study. The sample item is *I often feel depressed with the emergence of the novel pandemic of COVID-19*. The alpha reliability of this scale in this study was *0.868*.

Readiness for e-learning: To measure e-learning readiness, a 10-item scale of five dimensions developed by Hung et al. (2010) was used. The sample item includes, *I feel confident in using online tools to effectively communicate with others*. The alpha reliability of this scale in this study was *0.849*.

Academic performance: The 6-item scale of students' academic performance was measured. The scale is developed by Alamri et al. (2020). The sample item is "Has led to a better learning experience in online classes." The alpha reliability of the scale for this study was *0.890*.

Change-oriented leadership: The short version of change-oriented leadership was used to assess change-oriented leadership in times of COVID-19. The scale was

TABLE 11.1
Description of Sample and Characteristics

	Sample
Questionnaire emailed	345
Questionnaire returned	264
Response rate %	76.5%
Gender	
Male	204 (77%)
Female	60 (23%)
Education	
Bachelors	230 (87.12%)
Masters	34 (12.87%)
Other	–
Region	
Urban	140
Sub-urban	96
Rural	28

developed and used by King (2017). We used the scale with some modification in context of the study. The alpha reliability of this scale in this study was *0.727.*

11.4 DATA ANALYSIS AND FINDINGS

11.4.1 Measurement Model

The measurement model included psychological distress, student's readiness for e-learning, change-oriented leadership, and academic performance. We removed three items of e-learning readiness, two items of change-oriented leadership, one item of psychological distress, and two items of academic performance as the items did not meet the criteria of 0.7 thresholds (Hair et al., 2016). The scale reliability was tested through α values and composite reliability (CR). The convergent validity through average variance extracted (AVE) satisfies the criteria of a value higher than 0.5 (Hair et al., 2010). Tables 11.2 and 11.3 show the factor loadings along with the reliability and discriminant validity.

11.5 STRUCTURAL MODEL

The researchers used PLS-SEM in Smart PLS 3.3 to test the research model (Ringle et al., 2015). Particularly, to compare between-group differences of two countries, we used multi-group PLS analysis (Ringle, Wende, and Will, 2005) to test measurement and structural model analysis.

TABLE 11.2
Measurement Model Results

Construct	Loadings	Alpha	CR	AVE
	Psychological Distress			
PD2	0.872	0.868	0.910	0.716
PD3	0.834			
PD4	0.848			
PD5	0.829			
	Readiness for e-Learning			
RD1	0.824	0.849	0.882	0.518
RD2	0.686			
RD3	0.675			
RD4	0.690			
RD6	0.616			
RD8	0.806			
RD10	0.716			
	C-O Leadership			
COL1	0.957	0.727	0.828	0.711
COL2	0.711			
	Academic Performance			
AP1	0.906	0.890	0.923	0.750
AP2	0.847			
AP3	0.852			
AP4	0.857			

TABLE 11.3
Discriminant Validity

	AP	CL	PD	RD
AP	0.866			
COL	−0.034	0.843		
PD	−0.188	−0.007	0.028	
RD	0.234	−0.106	−0.270	0.720

11.5.1 Hypothesis Testing

The path coefficient of relationships, significance level, and R^2 to explain variance are important PLS elements for data analysis (Chin, 2010). The first hypothesis about the relationship between psychological distress and academic performance (H1) was analyzed through bootstrapping by running 1,000 subsamples. The analysis showed support for the relationship. The results are reported in Table 11.4.

TABLE 11.4
Path Coefficients

Relationships	B-value	Standard Deviation (STDEV)	T Statistics (|O/STDEV|)	*p* Values	Support for Hypothesis
PsyD -> AP	−0.135	0.060	2.243	0.025	H1 Supported
PsyD -> RD	−0.275	0.058	4.733	0.000	
RD -> AP	0.198	0.061	3.245	0.001	

TABLE 11.5
Specific Indirect Effects

	Pakistani Sample				
Relationships	Original Sample	*p*-value	Lower Threshold	Upper Threshold	Hypothesis
PsyD -> RD -> AP	−0.054	0.020	−0.098	−0.022	H2 Supported

TABLE 11.6
Total Indirect Effects

Relationships	Coefficients	*p*-value
PsyD -> AP	−0.054	0.006

The second hypothesis (H2) regarding the mediating effect of students' e-learning readiness in the relationship between psychological distress and academic performance was tested through indirect effects analysis (Zhao et al., 2010). The data analysis gave the total and specific indirect effects results. The results show that the hypotheses are supported by the data. The results for specific indirect effects and total indirect effects are shown in Tables 11.5 and 11.6.

The moderating effect of change-oriented leadership in the relationship between psychological distress and readiness (H3) was assessed through interaction effects in SmartPLS. The interaction between psychological distress and change-oriented leadership shows significant result, thus supporting the moderating effect of change-oriented leadership. Figure 11.2 shows the structural model of the study. Table 11.7 presents the results of moderation.

Finally, in order to test the moderated mediation of change-oriented leadership in affecting the indirect effect of psychological distress on academic performance through e-learning readiness at high and low values of change-oriented leadership (H4), bootstrap method in SPSS macro (Preacher et al., 2007) was used. The high (one standard deviation above the mean) and low values (one standard deviation below the mean) of change-oriented leadership were set. After running the test under model number 7 in PROCESS macro, the results show the support for hypothesized relationship.

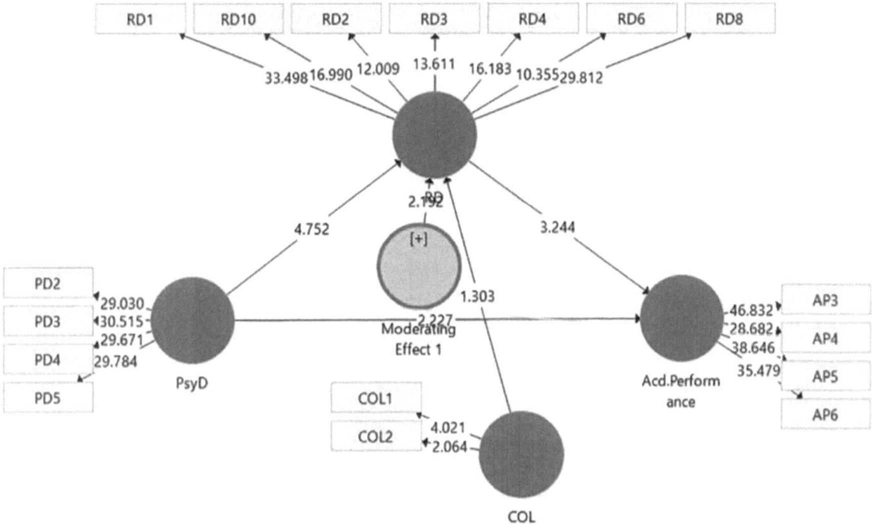

FIGURE 11.2 Structural model of study.

TABLE 11.7
Moderation Table

Relationships	B-Value	S.D	T Statistics	*p* Values	Support for Hypothesis
COL◊RD	0.083	0.064	1.312	0.190	H3 supported
Interaction effect PsyD × COL	0.143	0.064	2.212	0.027	

The moderated mediation was significant and stronger (–0.195) at high value of change-oriented leadership (CI ranging from – 0.062 to – 0.014). The moderated mediation was significant and weaker (–0.067) at low value of change-oriented leadership (CI ranging from –0.130 to –0.014). Thus, results of moderated mediation supported H4.

11.6 DISCUSSIONS AND CONTRIBUTIONS

This study investigated the impact of students' psychological distress during COVID-19 on their academic performance. Four relationships were hypothesized in this study. The data supported H1 about the negative relationship between students' psychological distress during COVID-19 and their academic performance. The novelty of e-learning due to the COVID-19 pandemic were majorly looked at as a hurdle by the many HEI's students because they were not ready for it. Within a short period, COVID-19 turned the world of HEIs upside down. At the individual level, students faced major depressive stress due to lack of social, emotional, and technical resources (Shakeel et al., 2020; Noor et al., 2020). All these factors added fuel to students' stress, which affected their performance.

The results also supported H2 about the mediating effect of e-learning readiness in the relationship between psychological distress and academic performance. The results shows evidence that the students expressed their discomfort in e-learning. For Pakistani HEIs, newly adopted technology-driven learning platforms, internet connectivity, virtual classrooms, digital study materials, mediated communication, missing peers, and social isolation added an enormous amount of stress toward student's academic life and dampened the e-learning readiness. The culmination of all the above demotivated students to get ready for e-learning (Qazi et al., 2020; Noor et al., 2020), which directly implies the negative effect on academic performance.

The change-oriented leadership's moderation in the relationship between psychological distress and e-learning readiness (H3) was also supported by the results. The significance thus reaffirms the importance of leadership in bringing positive change and fueling e-learning readiness among students. Leaders are the main figures in any organization and thus have significant power to bring change to tackle any challenge. COVID-19 motivated academic leaders to sustain and improve teaching and learning process through change-oriented behavior. It is possible because leaders have the position to provide resource support and infrastructure such as time, budget, knowledge, and support (Afshan and Serrano-Archimi, 2020; Sahibzada et al., 2020a, b).

The test of moderated mediation of change-oriented leadership (H4) was also supported by the results. These results shows that academic leaders through their change-oriented behavior can bring enormous changes in academic performance even in critical situation such as COVID-19.

There are three major contributions of the study. First, this is one among the recent studies that examines the psychological issues that students face during COVID-19 in developing countries. This investigation has provided an understanding of psychological issues of students caused by COVID-19-related factors and contributes to the recent studies in this context. The findings help us to understand how students in Pakistan are facing such issues and what could be the possible solutions to provide support to the students. Second, this study has been conducted on the student sample of HEIs in Pakistan. By studying the hypothesized relationships in Pakistan, which is a developing country and collectivistic society, the results provide an insight on how such countries are facing and tackling COVID-19, particularly in HEIs wherein the online learning does not usually take place. Third, change-oriented behavior in facilitating students' e-learning readiness and academic performance provide an understanding of academic leaders' role in HEIs to sustain the educational systems. The findings show that it is imperative to lead through change-oriented behavior to facilitate students' e-learning in HEIs to enhance their productive and innovative contributions to academic success. In conclusion, empirical support of hypothesized relationships contributes in extending the current knowledge of online education, academic leadership, and student's emotional and technical issues in affecting their performance.

11.6.1 Managerial Implications

The e-learning and COVID-19 mark the end of an era and the beginning of another. Every stakeholder in educational institutions has to work in students' best interest to tackle such a crisis in the future. This study's findings offer implications for the academicians, governing bodies of HEIs, and the government.

The COVID-19 pandemic is severe and destructive in nature that has largely affected the effectiveness of leadership strategies for student support system (Raaper and Brown, 2020). Therefore, heads of HEIs must understand the crisis fully and work toward delivering goods to the student's community. Their emotional, professional, and personal support is vital in selecting the best possible way forward to overcome any hurdle developed due to troubles like COVID-19. In particular, the technology, IT infrastructure support, training and development on e-learning platforms, applications, and learning the nuances stood out as major components in achieving confidence and compatibility for better teaching and learning experiences. There needs to be an immediate focus on students' psychological issues for a deeper understanding of college students' mental health development during the COVID-19 outbreak.

Developing countries like Pakistan need to focus on e-learning and blended learning pedagogy to engage learners. The HEIs need to introduce the students during their earlier days of higher education itself to e-learning applications and platforms, encouraging their enrolment.

11.6.2 Limitations and Future Directions

The study suffers from some limitations based on which we offer future directions. First, the data were collected from the students only. However, the multi-source data from teachers or parents could add value to the study's findings. Moreover, we have investigated the change-oriented leadership data from students only, which could mean we missed the possibility of not observing the significant role of change-oriented leadership in this study. Data from faculty, staff, and management could provide a complete picture of leaders in a crisis such as COVID-19. Second, the study is limited to Pakistan only. However, other developing countries could also be examined to see how the rest of the developing countries' academics have tackled the crisis of COVID-19. Finally, the study is cross-sectional, wherein the data has been collected at one period of time. Time lagged or longitudinal data could be value-added to see the behavioral changes over the time since the emergence of a pandemic.

REFERENCES

Adebola, O. O., Tsotetsi, C. T., & Omodan, B. I. (2020). Enhancing students' academic performance in university system: The perspective of supplemental instruction. *International Journal of Learning, Teaching and Educational Research, 19*(5), 217–230.

Afshan, G., & Serrano-Archimi, C. (2020). My supervisor supports me better than the past: integration of social and temporal comparison. *Journal of Economic and Administrative Sciences*. doi:10.1108/JEAS-07-2020-0128.

Agarwal, A., Agarwal, B., Harjule, P., & Agarwal, A. (2021). Mental health analysis of students in major cities of India during COVID-19. In: Khosla P.K., Mittal M., Sharma D., Goyal L.M. (Eds.). *Predictive and Preventive Measures for Covid-19 Pandemic* (pp. 51–67): Singapore: Springer.

Ahmad, A., Rahman, I., & Agarwal, M. (2020). Factors influencing mental health during COVID-19 outbreak: An exploratory survey among Indian population. *MedRxiv*. doi:10.17532/jhsci.2020.950.

Akinkugbe, A. A., Garcia, D. T., Smith, C. S., Brickhouse, T. H., & Mosavel, M. (2020). A descriptive pilot study of the immediate impacts of COVID-19 on dental and dental hygiene students' readiness and wellness. *Journal of Dental Education,* 85: 401–410.

Alam, A. (2020). Challenges and possibilities of online education during Covid–19.

Alamri, M. M., Almaiah, M. A., & Al-Rahmi, W. M. (2020). Social media applications affecting Students' academic performance: A model developed for sustainability in higher education. *Sustainability*, *12*(16), 6471.

Alderman, H., Orazem, P. F., & Paterno, E. M. (2001). School quality, school cost, and the public/private school choices of low-income households in Pakistan. *Journal of Human Resources*, *36*, 304–326.

Ali, M., Raza, S. A., Qazi, W., & Puah, C. H. (2018). Assessing e-learning system in higher education institutes. *Interactive Technology and Smart Education*. doi:10.1108/ITSE-02-2017-0012.

Alqahtani, A. Y., & Rajkhan, A. A. (2020). E-learning critical success factors during the covid-19 pandemic: A comprehensive analysis of e-learning managerial perspectives. *Education Sciences*, *10*(9), 216.

Aristovnik, A., Keržič, D., Ravšelj, D., Tomaževič, N., & Umek, L. (2020). Impacts of the COVID-19 pandemic on life of higher education students: A global perspective. *Sustainability*, *12*(20), 8438.

Asad, M. M., Hussain, N., Wadho, M., Khand, Z. H., & Churi, P. P. (2020). Integration of e-learning technologies for interactive teaching and learning process: An empirical study on higher education institutes of Pakistan. *Journal of Applied Research in Higher Education*, *13*, 649–663.

Aucejo, E. M., French, J., Araya, M. P. U., & Zafar, B. (2020). The impact of COVID-19 on student experiences and expectations: Evidence from a survey. *Journal of public economics*, *191*, 104271.

Bao, W. (2020). COVID-19 and online teaching in higher education: A case study of Peking University. *Human Behavior and Emerging Technologies*, *2*(2), 113–115.

Beaunoyer, E., Dupéré, S., & Guitton, M. J. (2020). COVID-19 and digital inequalities: Reciprocal impacts and mitigation strategies. *Computers in Human Behavior*, *111*, 106424.

Cao, W., Fang, Z., Hou, G., Han, M., Xu, X., Dong, J., & Zheng, J. (2020). The psychological impact of the COVID-19 epidemic on college students in China. *Psychiatry research*, *287*, 112934.

Chin, W.W. (2010). How to write up and report PLS analyses. In: Esposito Vinzi V., Chin W.W., Henseler J., & Wang H. (Eds.). *Handbook of Partial Least Squares* (pp. 655–690): Heidelberg: Springer.

Coccia, M. (2020). Factors determining the diffusion of COVID-19 and suggested strategy to prevent future accelerated viral infectivity similar to COVID. *Science of the Total Environment*, *729*, 138474.

Demir, O. (2015). The investigation of e-learning readiness of students and faculty members: Hacettepe University Faculty of Education example [Master Thesis]. *Ankara: Hacettepe University.*

Dever, C., & Justice, G. (2020, March 27). Coming together in crisis times, Inside Higher Ed. Retrieved from https://www.insidehighered.com/advice/2020/03/27/during-pandemic-crisis-faculty-and-administrators-must-move-beyond-their.

Dirani, K. M., Abadi, M., Alizadeh, A., Barhate, B., Garza, R. C., Gunasekara, N., … & Majzun, Z. (2020). Leadership competencies and the essential role of human resource development in times of crisis: A response to Covid-19 pandemic. *Human Resource Development International*, *23*(4), 380–394.

Eveland, T. J. (2020). Supporting first-generation college students: analyzing academic and social support's effects on academic performance. *Journal of Further and Higher Education*, *44*(8), 1039–1051.

Farooq, F., Rathore, F. A., & Mansoor, S. N. (2020). Challenges of online medical education in Pakistan during COVID-19 pandemic. *Journal of College of Physicians and Surgeons Pakistan*, *30*(6), 67–69.

Farrer, L. M., Gulliver, A., Bennett, K., Fassnacht, D. B., & Griffiths, K. M. (2016). Demographic and psychosocial predictors of major depression and generalised anxiety disorder in Australian university students. *BMC Psychiatry*, *16*(1), 1–9.

Fawzy, M., & Hamed, S. A. (2017). Prevalence of psychological stress, depression and anxiety among medical students in Egypt. *Psychiatry Research*, *255*, 186–194.

Fernandez, A. A., & Shaw, G. P. (2020). Academic leadership in a time of crisis: The coronavirus and COVID-19. *Journal of Leadership Studies*, *14*(1), 39–45.

Fernandez, A., & Shaw, G. (2020). Leadership in higher education in an era of adaptive challenges. In *International Technology, Education and Development Conference* (pp. 61–65). Valencia, Spain: IATED Academy.

Greenhow, C., Lewin, C., & Staudt Willet, K. B. (2020). The educational response to Covid-19 across two countries: A critical examination of initial digital pedagogy adoption. *Technology, Pedagogy and Education*, *30*, 1–19.

Hao, Y. (2016). Exploring undergraduates' perspectives and flipped learning readiness in their flipped classrooms. *Computers in Human Behavior*, *59*, 82–92.

Hasan, N., & Bao, Y. (2020). Impact of "e-Learning crack-up" perception on psychological distress among college students during COVID-19 pandemic: A mediating role of "fear of academic year loss". *Children and Youth Services Review*, *118*, 105355.

Hung, M. L., Chou, C., Chen, C. H., & Own, Z. Y. (2010). Learner readiness for online learning: Scale development and student perceptions. *Computers & Education*, *55*(3), 1080–1090.

Iqbal, M. J., & Ahmad, M. (2010). Enhancing quality of education through e-learning: The case study of Allama Iqbal Open University. *Turkish Online Journal of Distance Education, 11*, 84–97.

Izumi, T., Sukhwani, V., Surjan, A., & Shaw, R. (2020). Managing and responding to pandemics in higher educational institutions: Initial learning from COVID-19. *International Journal of Disaster Resilience in the Built Environment*, *12*, 51–66.

Jæger, M. M., & Blaabæk, E. H. (2020). Inequality in learning opportunities during Covid-19: Evidence from library takeout. *Research in Social Stratification and Mobility*, *68*, 100524.

Jena, P. K. (2020). Impact of Pandemic COVID-19 on Education in India. *International Journal of Research in Commerce and Management*, *5*(10), 7–11. doi:10.24941/ijcr.39209.07.2020.

Joshi, A., Vinay, M., & Bhaskar, P. (2020). Impact of coronavirus pandemic on the Indian education sector: Perspectives of teachers on online teaching and assessments. *Interactive Technology and Smart Education*. doi:10.1108/ITSE-06-2020-0087.

Kapasia, N., Paul, P., Roy, A., Saha, J., Zaveri, A., Mallick, R., ... & Chouhan, P. (2020). Impact of lockdown on learning status of undergraduate and postgraduate students during COVID-19 pandemic in West Bengal, India. *Children and Youth Services Review*, *116*, 105194.

Kaur, K., & Zoraini Wati, A. (2004). An assessment of e-learning readiness at Open University Malaysia, 1017–1022.

Khalil, R., Mansour, A. E., Fadda, W. A., Almisnid, K., Aldamegh, M., Al-Nafeesah, A., ... & Al-Wutayd, O. (2020). The sudden transition to synchronized online learning during the COVID-19 pandemic in Saudi Arabia: A qualitative study exploring medical students' perspectives. *BMC medical education*, *20*(1), 1–10.

Khan, K. S., Mamun, M. A., Griffiths, M. D., & Ullah, I. (2020). The mental health impact of the COVID-19 pandemic across different cohorts. *International Journal of Mental Health and Addiction*, *2020*, 1–7.

Khan, Z. H., & Abid, M. I. (2021). Distance learning in engineering education: Challenges and opportunities during COVID-19 pandemic crisis in Pakistan. *The International Journal of Electrical Engineering & Education*. doi:10.1177/0020720920988493.

King Jr., G. (2017). A Comparison Between Administrative Leadership in Higher Education and Change-Oriented Leadership in Banks.

Kotter, J. P. (2012). *Leading Change*. Boston, MA: Harvard Business Press.

Lathabhavan, R., & Griffiths, M. (2020). First case of student suicide in India due to the COVID-19 education crisis: a brief report and preventive measures. *Asian Journal of Psychiatry*.

Le Fevre, D. M., & Robinson, V. M. (2015). The interpersonal challenges of instructional leadership: Principals' effectiveness in conversations about performance issues. *Educational Administration Quarterly*, *51*(1), 58–95.

Lee, J. (2020). Mental health effects of school closures during COVID-1. *The Lancet Child & Adolescent Health*, 4(6), 421. doi:10.1016/S2352-4642(20)30109-7.

Majumdar, P., Biswas, A., & Sahu, S. (2020). COVID-19 pandemic and lockdown: cause of sleep disruption, depression, somatic pain, and increased screen exposure of office workers and students of India. *Chronobiology International*, *37*(8), 1191–1200.

Mamun, M. A., Sakib, N., Gozal, D., Bhuiyan, A. I., Hossain, S., Bodrud-Doza, M., ... & Pakpour, A. H. (2021). The COVID-19 pandemic and serious psychological consequences in Bangladesh: A population-based nationwide study. *Journal of Affective Disorders*, *279*, 462–472.

Nandy, M., Lodh, S., & Tang, A. (2020). Lessons from COVID-19 and a resilience model for higher education. *Industry and Higher Education*. doi:10.1177/0950422220962696.

Noor, S., Ali, M. N., & Husnine, S. M. (2020). Performance of online classes in Lahore, Pakistan During Covid-19. *Performance Improvement*, *59*(9), 33–42.

Odriozola-González, P., Planchuelo-Gómez, Á., Irurtia, M. J., & de Luis-García, R. (2020). Psychological effects of the COVID-19 outbreak and lockdown among students and workers of a Spanish university. *Psychiatry research*, *290*, 113108.

Paixão, G., Mills, C., McKimm, J., Hassanien, M. A., & Al-Hayani, A. A. (2020). Leadership in a crisis: doing things differently, doing different things. *British Journal of Hospital Medicine*, *81*(11), 1–9.

Pather, N., Blyth, P., Chapman, J. A., Dayal, M. R., Flack, N. A., Fogg, Q. A., ... & Lazarus, M. D. (2020). Forced disruption of anatomy education in Australia and New Zealand: An acute response to the Covid-19 pandemic. *Anatomical Sciences Education*, *13*(3), 284–300.

Phutela, N., & Dwivedi, S. (2020). A qualitative study of students' perspective on e-learning adoption in India. *Journal of Applied Research in Higher Education*, *12*, 545–559.

Preacher, K. J., Rucker, D. D., & Hayes, A. F. (2007). Addressing moderated mediation hypotheses: Theory, methods, and prescriptions. *Multivariate Behavioral Research*, *42*(1), 185–227.

Qazi, A., Naseer, K., Qazi, J., AlSalman, H., Naseem, U., Yang, S., ... & Gumaei, A. (2020). Conventional to online education during COVID-19 pandemic: Do develop and underdeveloped nations cope alike. *Children and Youth Services Review*, *119*, 105582.

Raaper, R., & Brown, C. (2020). The Covid-19 pandemic and the dissolution of the university campus: Implications for student support practice. *Journal of Professional Capital and Community*. doi:10.1108/JPCC-06-2020-0032.

Radha, R., Mahalakshmi, K., Kumar, V. S., & Saravanakumar, A. R. (2020). E-Learning during lockdown of Covid-19 pandemic: A global perspective. *International Journal of Control and Automation*, *13*(4), 1088–1099.

Rahiem, M. D. (2021). Remaining motivated despite the limitations: University students' learning propensity during the COVID-19 pandemic. *Children and Youth Services Review*, *120*, 105802.

Rehman, U., Shahnawaz, M. G., Khan, N. H., Kharshiing, K. D., Khursheed, M., Gupta, K., ... & Uniyal, R. (2020). Depression, anxiety and stress among Indians in times of Covid-19 lockdown. *Community Mental Health Journal, 57*, 1–7.

Riaz, M., Abid, M., & Bano, Z. (2021). Psychological problems in general population during covid-19 pandemic in Pakistan: Role of cognitive emotion regulation. *Annals of Medicine, 53*(1), 189–196.

Ringle, C. M., Wende, S., & Will, A. (2005). SmartPLS 2.0 (beta).

Ringle, C., Da Silva, D., & Bido, D. (2015). Structural equation modeling with the SmartPLS.

Roy, D., Tripathy, S., Kar, S., Sharma, N., Verma, S., & Kaushal, V. (2020). Study of knowledge, attitude, anxiety & perceived mental healthcare need in Indian population during COVID-19 pandemic. *Asian Journal of Psychiatry, 51*, 102083.

Sá, M. J., & Serpa, S. (2020). The COVID-19 pandemic as an opportunity to foster the sustainable development of teaching in higher education. *Sustainability, 12*(20), 8525.

Sahi, P. K., Mishra, D., & Singh, T. (2020). Medical education amid the COVID-19 pandemic. *Indian pediatrics, 57*(7), 652–657.

Sahibzada, U. F., Jianfeng, C., Latif, K. F., Shafait, Z., & Sahibzada, H. F. (2020a). Interpreting the impact of knowledge management processes on organizational performance in Chinese higher education: Mediating role of knowledge worker productivity. *Studies in Higher Education*, 1–18. doi:10.1080/03075079.2020.1793930.

Sahibzada, U. F., Jianfeng, C., Latif, K. F., Shah, S. A., & Sahibzada, H. F. (2020b). Refuelling knowledge management processes towards organisational performance: Mediating role of creative organisational learning. *Knowledge Management Research & Practice*, 1–13. doi:10.1080/14778238.2020.1787802.

Seraji, F., & Yarmohammadi, M. (2010). Preparation and readiness assessment validate incoming e-Learning courses. *Journal of Educational Measurement, 2*(32), 127–149.

Shahzad, A., Hassan, R., Aremu, A. Y., Hussain, A., & Lodhi, R. N. (2020). Effects of COVID-19 in E-learning on higher education institution students: The group comparison between male and female. *Quality & Quantity, 55*, 1–22.

Shakeel, S., Sabir, S., & Anwar, P. (2020). Mental health and academic performance of students during Covid-19 in Pakistan. *Mediterranean Journal of Basic and Applied Sciences (MJBAS), 4*(4), 114–123.

Son, C., Hegde, S., Smith, A., Wang, X., & Sasangohar, F. (2020). Effects of COVID-19 on college students' mental health in the United States: Interview survey study. *Journal of Medical Internet Research, 22*(9), e21279.

Spinelli, A., & Pellino, G. (2020). COVID-19 pandemic: Perspectives on an unfolding crisis. *The British Journal of Surgery*, 107, 785–787

Stephenson, J. (2001). *Teaching and Learning Online: New Pedagogies for New Technologies (Creating Success): Stephenson, J.* London: Kogan Page.

Temsah, M. H., Al-Sohime, F., Alamro, N., Al-Eyadhy, A., Al-Hasan, K., Jamal, A., ... & Somily, A. M. (2020). The psychological impact of COVID-19 pandemic on health care workers in a MERS-CoV endemic country. *Journal of Infection and Public Health, 13*(6), 877–882.

Thongsri, N., Shen, L., & Bao, Y. (2020). Investigating academic major differences in perception of computer self-efficacy and intention toward e-learning adoption in China. *Innovations in Education and Teaching International, 57*(5), 577–589.

Wang, X., Hegde, S., Son, C., Keller, B., Smith, A., & Sasangohar, F. (2020). Investigating mental health of US college students during the COVID-19 pandemic: Cross-sectional survey study. *Journal of Medical Internet Research, 22*(9), e22817.

Wei and Chou (2020) found that student's readiness mediated their satisfaction with the course.

Wei, H. C., & Chou, C. (2020). Online learning performance and satisfaction: do perceptions and readiness matter? *Distance Education, 41*(1), 48–69.

Whitelaw, S., Mamas, M. A., Topol, E., & Van Spall, H. G. (2020). Applications of digital technology in COVID-19 pandemic planning and response. *The Lancet Digital Health*, 2, E435–E440.

Yassin, A. A., Razak, N. A., Saeed, M. A., Al-Maliki, M. A. A., & Al-Habies, F. A. (2021). Psychological impact of the COVID-19 pandemic on local and international students in Malaysian universities. *Asian Education and Development Studies*. doi:10.1108/AEDS-05-2020-0098.

Yen, E. L. Y., & Mohamad, M. (2020). Utilising E-Learning to Assist Primary School ESL pupils in learning to spell during COVID-19 Pandemic: A literature review. *Creative Education*, *11*(8), 1223–1230.

Yilmaz, R. (2017). Exploring the role of e-learning readiness on student satisfaction and motivation in flipped classroom. *Computers in Human Behavior*, *70*, 251–260.

Zhao, X., Lynch Jr., J.G. and Chen, Q. (2010), "Reconsidering baron and Kenny: Myths and truths about mediation analysis. *Journal of Consumer Research*, *37*(2), 197–206. doi:10.1086/651257.

Zhou, Y., Zhao, J., & Zhang, J. (2020). Prediction of learners' dropout in E-learning based on the unusual behaviors. *Interactive Learning Environments*, 1–25. doi:10.1080/10494820.2020.1857788.

12 Digital Divide and Social Inequalities

Sociological Perspectives on Technology and Education

Al Karim Datoo
Sukkur IBA University

Fakhr-Un-Nisa Kurio
Sukkur IBA University

CONTENTS

12.1 INTRODUCTION

During the COVID-19 pandemic, the world has simultaneously been experiencing a 'double gesture' of social-psychological movement/experience. On the one hand, there is a slowing down of the physical and face-to-face interaction (including global travel due to restrictions), a kind of "'slowbalisation" phenomenon; and on the other, there is an experience of acceleration in the digitally mediated information and communication, dominating and redefining the 'social' a 'new normal, a kind of digital 'colonization', across the virtual-scopes. In the context of education too, one witnessed a spike in penetration of digital technologies in teaching and learning. Digital-based distanced learning (Online learning) has been championed as a tackler of the issue of educational disruption caused due to the global pandemic. The rise of the international discourse of the digital, galvanized by neo-liberal global capitalist market logic and interest, appears to be convincingly 'selling' digital technologies as "panacea" to cure all ills of the globally inflicted educational disruption. Such one size fits all 'solution' championing, overlooks varied and uneven political,

DOI: 10.1201/9781003143796-12

socio-economic, demographic, and infrastructure factors, as well as leads to the marginalization of many local educational sites and the learners.

This chapter contends that while there surely is a great potential and ability in the digital to serve many large-scale human/societal solutions, its wholesale one-size fits all 'borrowing' as a 'panacea' for educational access and continuity needs to be critically debated and scrutinized, especially in the context of education in the global south, In this respect, engaging in critical explorations around notions of the "digital divide" and "digital inequalities" (but not limited to these) that persist not only in the global south but also in many parts of the global north is vital. These apparent symptoms of the digital divide point to deeper layers of inequalities, economic and social, which have exacerbated further during/due to the pandemic. The following discussion engages with these concerns, based on a brief review of the literature.

12.2 TECHNOLOGY—A PANACEA TO ADDRESS GLOBAL EDUCATIONAL DISRUPTION?

The global pandemic COVID-19 has paralyzed many functions of human societies at the global scale. Educational institutions globally have suffered disruption where schools, colleges, and universities across the globe have had to experience closures. Amid these circumstances, educational organizations have been forced to transition from an offline to an online mode of teaching. While digital-based learning, before the pandemic was also part of education (in many contexts), during the pandemic the role of digital-based technologies has seen a spike in the scale of its implementation. Many organizations that have historically been reluctant to change were forced to adopt emerging technology as a result of this epidemic. Numerous colleges throughout the world recognizing the urgent need in this situation digitalized their operations. In the middle of this turmoil, online education is now presented as a panacea (Dhawan, 2020).

Overnight, traditional classrooms were turned into e-classrooms, implying that instructors had altered their entire instructional strategy for addressing current professional conditions and also to respond to changing conditions. At this challenging time, the main question is not whether online educational instructional approaches would deliver high-quality education but rather in what way educational institutions would be capable of implementing online learning on a massive scale (Carey, 2020). The three biggest challenges for online teaching are *distance, scale, and personalized teaching and learning*. Institutional solutions will merely support humanity in combating this epidemic (Liguori & Winkler, 2020).

Qualitative research conducted during the pandemic by the lead author of this discussion led to the observation that the shift to digital-based learning has been recognized as a marker of quality of education by the school stakeholders in rural Sindh, Pakistan. In this regard, the research observes that the very quality of education is being redefined through the lens of the digital. In this regard, the quality markers of an educational institution include (1) provider of adequate and robust digital infrastructure to facilitate IC-based learning in schools, (2) facilitate teachers in developing ICT-based pedagogic competencies for effective online teaching, and (3) provide ongoing professional development opportunities to upgrade their ICT-based e-pedagogies (Datoo, 2021).

12.3 GLOBAL ASPIRATIONS, LOCAL CONSTRAINS: CHALLENGES OF GOING DIGITAL

Going digital, while holding both promise and potential, seen from practical sites in many parts of the world, is one that is faced with several challenges. In this regard, Aung and Khaing (2015) highlighted many obstacles in implementing an e-learning scheme some of which are categorized as (1) *technical challenges*, (2) *human challenges*, (3) *cultural challenges*, and (4) *course-related challenges*. These issues vary greatly from context to context. The main barriers to the digital-learning system approval in many nations were a lack of ICT expertise, a lack of network connectivity, and a lack of content creation (Aung & Khaing, 2015). Kanwal and Rehman (2017) described the key problems that impeded the successful implementation of an e-learning system in Pakistan; these were system features, internet experience, and machine self-efficacy. Also, certain factors such as Computer Anxiety (CA) and the Enjoyment of e-learning Device Use are challenges at the personal level.

Moreover, Almaiah et al. (2016) pronounced e-learning use and acceptance by operators as a difficult problem for many institutions of higher education in both developed and developing nations; however, in developing countries, students' willingness to consider and use the e-learning scheme is expected to be less of a problem. According to Eltahir (2019), the complexities of implementing an e-learning structure is due to the technological gap with developing nations. He stated that having a test center fitted with the newest informative and learning digital equipment and also secure internet access is still a fantasy for many institutions. This issue gets more complex when it comes to rural parts of many developing countries.

According to a United Nations Educational, Scientific and Cultural Organization study of developing countries' educators, "the majority of African educators stated that a shortage of sufficient hardware and software, as well as poor Internet connectivity, were significant obstacles to using computers in instruction" (Eltahir, 2019). According to the survey, many developing-country schools have an 80:1 student-teacher ratio and often contend with 20 or more electronic devices for the whole classroom. Furthermore, due to a lack of preparation and professional advancement in integrating technology into teaching, most teachers in developed countries lack the necessary skills needed to use advanced instructional technology resources (Eltahir, 2019).

Based data from the UNESCO Institute for Statistics and the International Telecommunication Union, it was shown that – "some 826 million students – kept out of the classroom by the COVID-19 pandemic, do not have access to a household computer and 43% (706 million) have no internet at home, at a time when digitally-based distance learning is used to ensure educational continuity in the vast majority of countries". These disparities are more prevalent in low-income countries. For example, in sub-Saharan Africa, 89% of learners do not have access to household computers, and 82% lack internet access. Furthermore, while mobile phones can enable access to information, around 56 million learners live in locations not served by mobile networks, almost half in sub-Saharan Africa (the same report informed). The economic and digital disparities point to questions of inequalities that are historically developed in many parts of the world. It is acknowledged that "while efforts to provide connectivity to all must be multiplied, we now know that continued teaching and learning cannot be limited to online means. To lessen already existing inequalities, we must also support other alternatives including the use of community radio and television broadcasts, and creativity in all ways of learning. These are solutions we are addressing with our Global Coalition partners", stated Audrey Azoulay, UNESCO Director-General.

12.4 DIGITAL DIVIDE

Many countries are facing technological 'gaps' due to internet connectivity, digital learning, and cultural/ethnic barriers. The issue of the digital divide is a complex one, but often gets described in terms of a single factor, typically as an issue of access and connectivity, resulting in a distorted picture of the issue. The issue of the digital gap is complex, involving spatial stratification of social class and race, and it affects both urban and rural areas. According to Scheerder et al. (2017), the 'digital gap' has been a concern for social scientists since the introduction of the internet. In this respect, in the literature there are three degrees of the digital divide. These include (1) connection to computers and the internet (first level), (2) the skills and use of technology for pedagogic purposes (second level), and (3) varied impact of technology access, expertise, and use on different outcomes (such as work, education, social and political connectedness, and health).

The connectivity varies across social classes and spatiality in the content affordability of computers, applications, subscriptions, and secondary facilities, including printers and external hard drives (Van Deursen, & Helsper, 2015). In many parts of the less developed countries, internet access is also limited and unequal (Rotondi et al., 2020). Moreover, the rural areas face an additional challenge when it comes to digital connectivity, as many rural areas still lack Internet connectivity (even electricity in many contexts), and those who still have connections use old dial-up modems, putting them at a serious technological disadvantage over Internet use (Haythornthwaite, 2007).

These views on the digital divide concentrate on technology users, analyzing what they have access to, possess the relevant expertise to use technology, and draw the gains of all of these in other domains of their lives. However, it should also be noted that the degree to which digital disparities arise through schooling is often influenced by the educational environment in which students are learning: classrooms. Whereas the accessible infrastructure has been categorized as an entry method (Van de Werfhorst, Kessenich, & Geven, 2020), likewise schools' technology for providing digital education can vary (e.g., by the availability of computers in the school). The contextual framework for digital learning then can differ not only in terms of the technology required for access to digital learning but also in terms of the skills required of teachers to provide the education required in an online environment. It could be added that another important dimension of the digital divide debate is paying attention to the capabilities available in the sense in which the students can read and comprehend. This aspect stresses that the results of internet usage are determined not just by an individual's preparation and capabilities but rather by the degree to which external parties build conditions conducive to the achievement of personal objectives (Van de Werfhorst et al., 2020), pointing somewhat to pedagogic framing to engage learners.

12.5 DYNAMICS OF DIGITAL DIVIDE: SOCIAL INEQUALITIES

The study of Chou and Shieh (2011) expressed that in a world of information, one way to measure disparity is across the digital divide. "Digital divide" refers to the gap in accessibility to telecommunications along with Internet usage between people, families, enterprises, and geographic areas at various socioeconomic levels (OECD). There are three aspects of the digital divide, according to academics: (1) ICT access, (2) ICT use, and (3) ICT applications. Physical access was traditionally emphasized in previous digital divide studies. Since 2002, however, there is a growing number of scholars advocating for moving "beyond entry" level considerations and focusing on social, psychological, and cultural factors of the technology–society interface. The majority of research on the digital divide has concentrated on its socio-demographic correlates and consumers' differing access to information and communication technology. Age, ethnicity, and education are all key variables in the digital divide, according to the research findings.

Additionally, gender may be an especially important variable to the perceptions/experiences of technophobia because coding is seen as a "masculine" practice (Chou & Shieh, 2011). Furthermore, age differences, in addition to gender differences, are thought to influence people's computer-related behaviors. According to educational gerontology research, the older a person gets, the less computer experience and interest they are likely to have (Chou & Shieh, 2011).

In addition to gender and age, Bozionelos (2004) articulated socio-economic considerations as having an influential link to the digital divide. The digital consumers' socio-economic status has a strong positive correlation with computing experience and an indirect negative correlation with computer anxiety. According to Chou and Shieh (2011), changes in computer usage are positively linked to changes in college graduates' employment share but negatively related to high school graduates' employment share. The digital divide is also shaped by other socio-economic influences. According to some studies, individuals with higher incomes use computers rather more than those in the middle- and lower-income groups (Chou & Shieh, 2011). Furthermore, Tien and Fu (2008) discovered no major discrepancies in terms of ethnic and socio-economic family history correlates in forecasting the different reasons for using machines. Nevertheless, gender/sex, socioeconomic family history, and academic majors and interests have an impact on computer skills and awareness.

It could be observed that the assumptions about the digital divide have shifted throughout the debate. Primarily, the relationship between ICTs and growth was thought to be linear and deterministic, as if having access to technologies led to social progress automatically. However, the notion of the digital divide became more nuanced as a result of subsequent intellectual and political discussions, especially when issues of technology and education are looked at from sociological perspectives. From a socio-technical standpoint, it is critical to understand the issue of digital divides in their geographic, communal, and chronological settings and in terms of the expectations and needs of different social groups and sectors, taking into account divides related to not only hardware and software access but also information, expertise, the skills, and environment required to engage technologies for learning.

To sum up, the concept of the digital divide can be categorized into and analyzed according to three dimensions: (1) the economic component of the digital divide, as access to equipment and communication, (2), cultural dimension, that is cultural and educational resources that influence ICT appropriation capability, and (3) analyzing "how" and "why" various consumers use technology, especially examining possibilities and/or difficulties faced by social groups in using digital technologies, transforming their situations, and improving their living conditions. In this context, digital skills and access are seen as important but not adequate criteria for ICT adoption (Larghi et al., 2015).

12.6 DIGITAL INEQUALITIES AND PANDEMIC: CONCLUDING REMARKS

As schools and other educational institutions move fast going forward with digital-based learning, the policymakers and educational leaders should not forget the social ontologies having digital divides/gaps that exacerbate already persisting digital and social inequalities. The discourse and approach of educational and developmental recovery needs to be sensitive and inclusive to these disparities that are likely to remove many young learners from the privilege and right to learn. Beckman et al. (2018) demonstrated that not all young people are the well-connected, technologically savvy, 'digital natives' that the discourse surrounding young people and technology will have us assume.

Furthermore, Williamson and colleagues (2020) reckon technology as a dynamic and social-cultural artifact and not a neutral force that merely does well when people have access to it. As discussed, digital-based learning is a complex phenomenon and involves a host of critical debates and considerations (as demonstrated through discussion of the digital divide), beyond the issue of internet access.

In the context of the pandemic, the internet will provide access to teachers, learning, and social resources by bringing technology into homes that are already quite likely to be struggling financially, and socio-psychologically, who have suffered even more since the COVID-19 epidemic. The same internet, on the other hand, offers payday loans and casino firms with better access to families that are still suffering financially, and data traders with more knowledge that could have a detrimental effect on the families' future. The digital therefore offers both possibilities and challenges. These challenges, along with the issues often seen with 'EdTech' firms and the types of technical education on wholesale offer these days necessitates a critical new approach to coping with issues of digital/inequality in such a manner that those already suffering and marginalized may not get further alienated and negatively impacted. Furthermore, the lack of digital access is mostly due to a lack of social and financial capital (a result of deep-seated socio-economic disparities in societies). These cultural and economic realities cannot be changed just by a laptop program only. Indeed, when the pandemic spreads, more young people and their communities may face financial deprivation, and societal gaps between Haves and Have-nots will possibly expand (Sancho-Gil, Rivera-Vargas, and Miño-Puigcercós, 2020).

There is an urgent need for critical reflection and action to rethink the whole issue of educational access, participation, and continuity to one that is sensitive and inclusive to the needs and social realities of the learners. Digital technologies are going to play a dominant role in shaping the future of education, both during and beyond this current pandemic. While technology offers a great promise, it should not be embraced uncritically as a wholesale 'solution'. The future role and scale of digital-based technologies in education need to be examined through concerns around the digital divide, digital inequalities, and broader social inequalities, which alienate the poor and the marginalized.

In this regard, paying attention to varied critical geographies in education, as well as applying a lens of sociology of technology and education can offer value in showing possibilities and limitations of digital technologies in addressing educational access and participation in the global South.

REFERENCES

Almaiah, M. A., Jalil, M. A., & Man, M. (2016). Preliminary study for exploring the major problems and activities of mobile learning system: A case study of Jordan. *Journal of Theoretical & Applied Information Technology, 93*(2), 580–594.

Aung, T. N., & Khaing, S. S. (2015, August). Challenges of implementing e-learning in developing countries: A review. In: Zin T., Lin J.W., Pan J.S., Tin P., & Yokota M. (Eds.). *International Conference on Genetic and Evolutionary Computing* (pp. 405–411): Cham: Springer.

Beckman, K., Apps, T., Bennett, S., & Lockyer, L. (2018). Conceptualising technology practice in education using Bourdieu's sociology. *Learning, Media and Technology, 43*(2), 197–210.

Bozionelos, N. (2004). Socio-economic background and computer use: The role of computer anxiety and computer experience in their relationship. *International Journal of Human-Computer Studies, 61*(5), 725–746.

Carey, K. (2020). Everybody ready for the big migration to online college? Actually, No. *The New York Times, 13.*

Chou, J. R., & Shieh, C. J. (2011). The digital divide in the Taiwanese unemployed adult population. *Scientific Research and Essays, 6*(7), 1514–1521.

Datoo, A. (2021). Contextualising the discourse on "Quality" in Education across Community Colleges, Sukkur-IBA University, Sindh, Pakistan (Unpublished report).

Dhawan, S. (2020). Online learning: A panacea in the time of the COVID-19 crisis. *Journal of Educational Technology Systems, 49*(1), 5–22.

Eltahir, M. E. (2019). E-learning in developing countries: Is it a panacea? A case study of Sudan. *IEEE Access, 7*, 97784–97792.

Haythornthwaite, C. (2007). Digital divide and e-learning. *The Sage Handbook of e-learning Research*, 97–118.

Kanwal, F., & Rehman, M. (2017). Factors affecting e-learning adoption in developing countries–empirical evidence from Pakistan's higher education sector. *IEEE Access, 5*, 10968–10978.

Larghi, S. B., Lemus, M., Moguillansky, M., & Welschinger, N. (2015). Digital and social inequalities: A qualitative assessment of the impact of the connecting equality program on Argentinean youth. *The Electronic Journal of Information Systems in Developing Countries, 69*(1), 1–20.

Liguori, E., & Winkler, C. (2020). From offline to online: Challenges and opportunities for entrepreneurship education following the COVID-19 pandemic.

Rotondi, V., Kashyap, R., Pesando, L. M., Spinelli, S., & Billari, F. C. (2020). Leveraging mobile phones to attain sustainable development. *Proceedings of the National Academy of Sciences, 117*(24), 13413–13420.

Sancho-Gil, J. M., Rivera-Vargas, P., & Miño-Puigcercós, R. (2020). Moving beyond the predictable failure of Ed-Tech initiatives. *Learning, Media and Technology, 45*(1), 61–75.

Scheerder, A., van Deursen, A., & van Dijk, J. (2017). Determinants of Internet skills, uses, and outcomes. A systematic review of the second-and third-level digital divide. *Telematics and Informatics, 34*(8), 1607–1624.

Tien, F. F., & Fu, T. T. (2008). The correlates of the digital divide and their impact on college student learning. *Computers & Education, 50*(1), 421–436.

Van de Werfhorst, H., Kessenich, E., & Geven, S. (2020). The Digital Divide in Online Education. Inequality in Digital Preparedness of Students and Schools before the Start of the COVID-19 Pandemic.

Van Deursen, A. J., & Helsper, E. J. (2015). The third-level digital divide: Who benefits most from being online? In *Communication and Information Technologies Annual.* Emerald Group Publishing Limited.

Williamson, B., Eynon, R., & Potter, J. (2020). Pandemic politics, pedagogies, and practices: Digital technologies and distance education during the coronavirus emergency.

13 Engineering Students' Motivation Toward Education 4.0

A Conceptual Framework for Rural Sindh

Hammad Siddiqui, Rameen Butt, Raheel Ahmed Soomro, Muhammad Mujtaba Asad, and Fahad Sherwani
Sukkur IBA University

CONTENTS

DOI: 10.1201/9781003143796-13

13.1 INTRODUCTION

Industrial revolution and information communication and technologies (ICTs) have transformed classical society to a digital society. They have completely changed the way of living (Puncreobutr, 2016). As per the need of this society, ICT-based pedagogies have been implemented in schools, colleges, and universities. Moreover, technology-based pedagogies have focused and enhanced students' life skills, critical thinking skills, creative thinking, problem solving skills, logical and reasoning skills, and collaboration skills. In addition to this, ICT-based pedagogies help students to become an independent learner (Al-Hariri & Al-Hattami, 2017). Previous research studies show that integration of ICT tools in education systems, like internet, laptops, and smart computers, has increased students' achievements ratio than traditional ones (Korpelainen, 2011). Previous research studies also show that ICT-based tools have increased students' motivation toward learning new skills. Moreover, Steve Johnson has written in his book "Everything bad is good for you" that technology makes us smarter and more intelligent people in the way we receive, analyze, and process the knowledge (Johnson, 2006). That is why the new concept of Industrial revolution 4.0 was launched. It is all about technology and innovation. It is defined as a blend of human and technology. It includes cyber-physical system, cloud storage, internet of things (IoT), wireless robots, etc. (Rojko, 2017). In countries like Germany, Malaysia, and Singapore, research has been done on industry 4.0 and its importance and implementation. It has even been successfully implemented in these countries (Bahrin et al., 2016). In order to meet the requirements of technology-based industry 4.0, they introduced a new concept, Education 4.0. It is defined as school of thoughts, which motivates students to think untraditionally. It includes technology-based tools and devices that help learners to develop abilities and learn new skills in order to compete in the industry 4.0. Moreover, it focuses on both students' content knowledge and skills for entire life (Halili, 2019). The Model of Education 4.0 is also successfully implemented in schools, colleges, and universities of Malaysia. That is why the core purpose of this research study is to measure the level of motivation toward Education 4.0 among Pre-Engineering and Premedical educators of community colleges in rural Sindh. A comparative study is done to differentiate the level of motivation among Pre-Engineering and Premedical educators of community colleges in rural Sindh Pakistan.

13.2 PROBLEM BACKGROUND

In the modern era, the economy is undergoing prodigious transformation. The 4th industry revolution brings the wide array of information and communication technologies that are changing and impacting the daily lives and creating the digital society (Puncreobutr, 2016). The technological changes and innovation in the industrial sectors are known as industrial revolution (Benešová & Tupa, 2017). These revolutions started from the mechanization (first industrial revolution), use of electrical energy (second industrial revolution), and electronics and automation (third industrial revolution) to the current fourth industrial revolution (IR 4.0) that is based on the digitalization and robotics (Benešová & Tupa, 2017). Industrial revolution 4.0 affects all the domains of humans' social and personal lives. It affected the education as well and demands more skill and knowledge to overcome the needs of current industrial needs.

Moreover, Dunwill (2016) says that the improvement and advancement of technologies keep on changing and transforming the teaching pedagogical methods and situation of the learning processes. Lineways (2017) argues that a rise in the utilization of ICTs in education with the incorporation of technology to the curriculum proves to be a milestone in the progress and achievements of the learners. Moreover, ICTs have proved to be an effective mode to enhance the learning of the students as well as the pedagogical practices of a teacher. The era of rapid technological revolution is now enlivening the educational planning, so it is necessary to add ICT in the education (Wasif et al., 2012). In addition to these points, Lase (2019) argues that teachers must have the abilities and expertise to adapt to or foster new technologies and global changes. Furthermore, Sudlow (2019) emphasized on the changing of old literacy (reading, writing, and mathematics) with the new data literacy (ability to read, analyze, and use information from the data), technology literacy (understanding the system of mechanics and technology), and human resources literacy (ability to interact well). Further, Hussin (2018) argued that Education 4.0 is well response to the needs of Industrial Revolution 4.0 where technology and human interact each other to take the benefit from new emerging opportunities in future. including Machine learning and Artificial intelligence. Fisk (2017) explicates that "the new vision of learning promotes learners to learn not only skills and knowledge that are needed but also to identify the source to learn these skills and knowledge." Education 4.0 meets the need of the new dynamic, rapidly changing world, which has transformed from an agricultural society to an industrial society (Puncreobutr, 2016). This education system needs to carry and speed up the new and experienced workforce with innovative and creative suggestions of industry 4.0 to accelerate the production (Mourtzis et al., 2018).

Fisk (2017) implies nine trends of Education 4.0, it is also cited by Hussin (2018) and Lase (2019), and this is shown in Table 13.1. In addition, many countries are now moving toward the knowledge-based economy that is directly linked with the fourth industrial revolution. In Pakistan, especially in Sindh, the technological use and implementation is limited due to the lack of resources. Whereas 60% of Pakistan's population is younger than 25 years, if the education system isn't providing the adequate opportunities for them, then social and political problems and consequences

TABLE 13.1
Fisk's (2017) Nine Trends of Education 4.0

S#	Trends
1.	Learning can take place anywhere, any time. E-learning tools offer great opportunity for remote and self-paced learning.
2.	Learning will be personalized to individual students.
3.	Choice how individual students want to learn.
4.	Students will be experienced about more project-based learning.
5.	Students will be more involved in hands-on learning, such as field trips, internships, outdoor projects and camps, etc.
6.	Students will be exposed to data interpretation in which they are required to apply their theoretical knowledge to numbers and use their reasoning skills to make interpretations based on logic and trends from given sets of data.
7.	Students will be assessed differently; the conventional way of assessing students may become inappropriate and inadequate.
8.	Students' opinion will become valuable in designing and updating curriculum.
9.	Students will become more independent in their learning; thus, teacher role will become that of the facilitator who only guides students in the learning processes.

will arise (Nishat & Michael, 2019). Pakistan's education system is living in the age of industrial revolution 2.0 and world is competing with IR 4.0. Thus, Pakistan and especially Sindh has to necessarily take more constructive steps toward the sustainable knowledge-based economy where individuals can enhance their skills for combating the rapid digitalization of the society.

13.3 PROBLEM STATEMENT

Pakistan is far away from the new trends in industry and education because of lack of resources and awareness. Pakistan is living in the second industrial revolution and world is currently exploring the fourth industrial revolution that causes to shift the industry from Pakistan to those areas that are technological well developed (Dar, 2017)(Dar, 2017) First, Pakistan is lacking in industrial and educational development as compared to other under developing countries due to limited resources, unsustainable educational system, political instability, and corruption. Whereas, in this advanced era of industrial revolution 4.0 and technological development with the integration of artificial intelligence and internet of things (IoTs), still there is sheer need of effective techno-pedagogical framework. Therefore, educators and stakeholders are not motivated to effectively embrace Education 4.0 which is retrieved from IR 4.0 in their pedagogical strategies and lesson plans.

Second, the knowledge-based economy policy of Pakistan will be implemented throughout the country to accomplish the objectives of Pakistan Vision 2025. additionally the digital Pakistan policy clearly stated that it is necessary to collaborate with relevant stakeholders, including International Universities, to advance e-learning for the educational and other strategic sectors to improve our knowledge economy

(Ministry of Information, Broadcasting, 2018). Finally, teachers are not ready to accept the rapid inclusiveness of ICT. There was less use of technology in classroom like showing videos and other content-related material (Makrides, 2019).

13.4 RESEARCH OBJECTIVES

In this study, the following research objectives have been designed to know about the level of motivation of educators of Community Colleges and Schools of rural Sindh toward the integration of Education 4.0.

- To assess the difference between the level of motivation of Premedical and Pre-Engineering educators of Community Colleges and Schools of rural Sindh toward the integration Education 4.0.
  a. To analyze the level of motivation toward Education 4.0 among Pre-Engineering educators of Community Colleges and Schools in rural Sindh, Pakistan.
  b. To identify the level of motivation for the integration of Education 4.0 among Premedical educators of Community Colleges and Schools in rural Sindh, Pakistan.

13.5 RESEARCH QUESTION

Based on the proposed objectives of this study, the following questions have been developed to know about the level of motivation of educators of Community Colleges and Schools in rural Sindh toward the integration of Education 4.0.

- What is the difference between the level of motivation of Pre-Engineering and Premedical educators of Community Colleges and Schools of rural Sindh toward the integration of Education 4.0?
  a. What is the level of motivation toward Education 4.0 among Pre-Engineering educators of Community Colleges and Schools in rural Sindh, Pakistan?
  b. What is the level of motivation toward Education 4.0 among Premedical educators of Community Colleges and Schools in rural Sindh, Pakistan?

13.6 RESEARCH HYPOTHESIS

Based on the research objectives and research questions, the following null and alternative hypotheses were designed to know about the level of motivation toward Education 4.0 among Pre-Engineering and Premedical educators of Community Colleges and Schools in rural Sindh, Pakistan.

13.6.1 Null Hypothesis

There is no significant difference between the mean score of level of motivation toward Education 4.0 among Pre-Engineering and Premedical educators of Community Colleges and schools in rural Sindh, Pakistan.

13.6.2 Alternative Hypothesis

There is a significant difference between the mean score of level of motivation toward Education 4.0 among Pre-Engineering and Premedical educators of Community Colleges and schools in rural Sindh, Pakistan.

13.7 SCOPE OF RESEARCH

As the Government of Pakistan in 2019 has declared the knowledge economy policy, there will be a paradigm shift toward the technology in educational system. In this regard, this research provides a notion of how much the teachers are motivated toward implementation of technology-based method of teaching and learning—Education 4.0. This study tends to provide some interesting afterthoughts of Education 4.0 in teaching and learning process for the upcoming years.

13.8 APPROACH OF RESEARCH

In this study, sequential explanatory research design approach will be adopted in which 100 educators will be randomly selected for quantitative research (survey), while 16 educators from Premedical and Pre-Engineering domains will be respondents of qualitative research (structural interviews) from Community Colleges and Schools of rural Sindh, Pakistan.

13.9 CONCEPTUAL FRAMEWORK

The conceptual framework has been designed to know the level of motivation toward Education 4.0 among Pre-Engineering and Premedical Educators of Community Colleges and Schools in rural Sindh, Pakistan, according to the factors determined through literature review. It reveals that there are three factors that affect the level of motivation toward Education 4.0. The first factor is Human Factor that includes Educators' self-efficacy toward Education 4.0. The second factor is Intrinsic Value that includes the integration of Education 4.0. The educators name the third factor as Influencing Factor that includes frequent use of ICTs. All these factors are supported with the theory of motivation called as "Theory of Self-Efficacy," shown in Figure 13.1.

13.10 LITERATURE REVIEW

In this chapter, we reviewed related literature regarding our topic "Motivational level of educators toward Education 4.0" to help us define the focus of our study and develop an outline for the gathering and analyzing the data. This literature review is about ICT and digital skills in education, motivational theories regarding use of ICT, brief introduction of industrial revolutions (IR 1.0, IR 2.0, IR 3.0, and IR 4.0), Educational revolutions (Education 1.0 to Education 4.0), and review of government policies to depict the picture of Education 4.0 scenario in Pakistan.

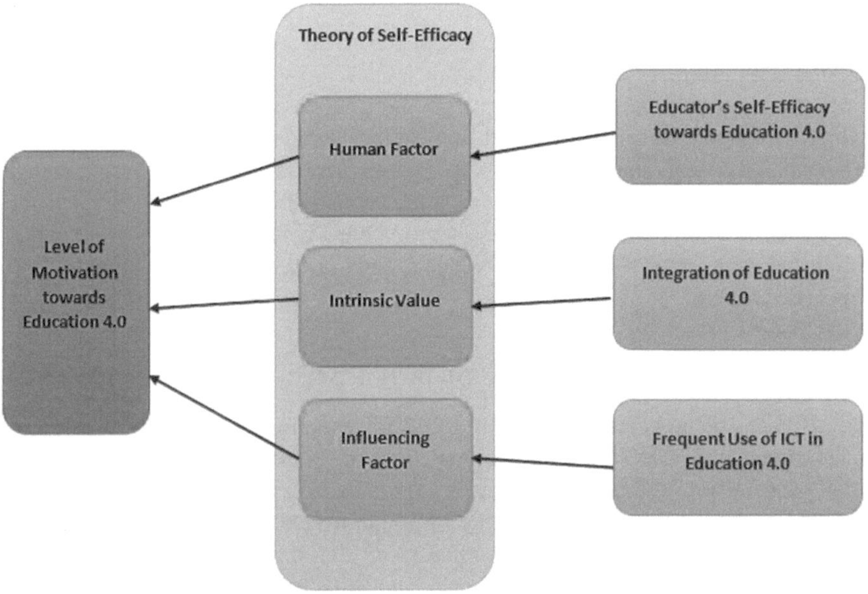

FIGURE 13.1 Conceptual framework.

13.10.1 ICT and Digital Skills in Education

ICTs merely focus on communication technologies (Christiansen, 2010). It is identified as technologies that develops to manage and communicate information from one place to another all over the world. It also holds the technologies that are used to store and save information and then retrieve or recall that information later when it is required (Techreviews, 2019). ICTs have proved to be an effective tool to enhance the learning as well as in pedagogical practices of the teachers. Lineways (2017) advocated that incorporation of technologies in the curriculum proved to be a milestone in the progress and achievement of the learners. Moreover, the modern technologies have potential to aid all through in curriculum and offer excessive opportunities for effective communication between teachers and students (Rahim Sajid, 2013). Wasif et al. (2012) endorsed that this is the era of modern technologies. Therefore, to enliven planning, it is important to incorporate ICT in all educational fields. Furthermore, the world is changing day by day, the need of technologies to communicate are also changing, and ICT is a backbone of this transitional world (Lemke, 2002). Farr and Murray (2016) argued that in a scenario of rapid developments, the world required new and furnished skills, and educators of this world should train adults as well as children on how to pilot, navigate, and direct the progressive digital world.

The term 21st-century skills also refer to skills that are needed for education and current economical workstations (van Laar et al., 2020). The Partnership for 21st Century Skills (P21, 2007) is a joint government–corporate organization that lists three types of skills: learning skills (creativity and innovation; critical thinking and problem solving; communication and collaboration), literacy skills (information literacy;

media literacy; ICT literacy), and life skills (flexibility and adaptability; initiative and self-direction; social and cross-cultural skills; productivity and accountability; leadership and responsibility). As the world is now connected with a network and internet connects and provide widespread place to get information easily, most of the researchers are agreed upon to use the internet in meaningful ways; thus, users must develop sufficient digital skills (Jenkins et al., 2009; Mossberger et al., 2003). As the digital age is related with digital tools, they are constantly changing the way teachers teach and students learn. Teachers need to keep up with evolving technology, know what digital tools and skills are best suited to their students, and use them effectively in their classrooms (Birla, 2018).

13.10.2 Motivation for Integration of ICT in Education

Various types of tools have been introduced for enabling the process of teaching and learning; nonetheless, the motivation to utilize these tools in system of education especially in the classroom varies from educator to educator (Schulz et al., 2015). Moreover, they also indicate the three main factors that affect the motivation of teachers to use ICTs. First is human factor that includes values and attitudes (Player-Koro, 2012), confidence, proficiency and skills to utilize ICTs (Ounis, 2016), and opinions that are held by the educators about ICT tools (Mahdum et al., 2019). Second is intrinsic values that holds satisfaction, level of interest, and factor of joy and happiness by using ICTs among the educators (Schulz et al., 2015). The last one is influencing factors including adaptability, meeting learning requirement, usability, and level of interaction (Schulz et al., 2015). Moreover, Mahdum et al. (2019) suggested that educators' overall motivation to use ICT depends upon two factors, internal factors such as educators own interest, experience, and readiness and external factors such as availability of resources, incentive, and job requirements.

13.10.3 Theories of Motivation of ICT

According to Zhang (2008), motivational theories explain the process in which behaviors give its whole energy including strength, intensity, and persistence and direction.

13.10.3.1 Theory of Self-efficacy

This social learning theory of motivation was pioneer by Albert Bandura in 1977. The term Self-efficacy is well defined as a person's confidence and belief in one's ability to accomplish a given task efficaciously. It controls and modifies the human behavior (Yusuf, 2011). Dullas (2018) quoted in his research that Albert argued that one's beliefs in personal self-efficacy influences his personal choices, motivation, and interest in accomplishing the tasks. Self-efficacy theory explains that human actions and success depends upon how deep the interaction is between the individual's thought and given task. Moreover, Yusuf (2011) also indicated that if an individual has superficial sense of self-efficacy, he will possess the negative thought and take that task as a threat; however, individuals with great sense of self-efficacy will possess the positive thoughts and take task as an opportunity. Furthermore, Albert

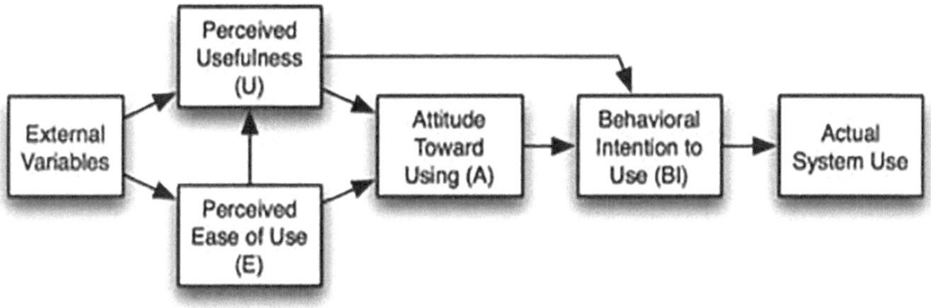

FIGURE 13.2 Technology acceptance model (TAM).

Bandura explained that there are four main sources of self-efficacy: (1) past performance, (2) vicarious experience, (3) verbal persuasion, and (4) emotional cues (Bandura, 2010).

13.10.3.2 Technology Acceptance Model (TAM)

Davis has developed the Technology Acceptance Model (TAM) in 1989. It has been extensively used for determining one's acceptance toward the ICTs (King & He, 2006). It has been used to define one's behavior to the utilization of technology. There are two key aspects in determining the technology acceptance: perceived usefulness and perceived ease of use. These two aspects shape the attitude and behavioral intents of an individual as well as the use of actual system by the individual (Surendran, 2012). Figure 13.2 shows TAM modal.

13.10.4 Industrial Revolution (IR)

Industrial revolution was a period during the late 1700s and early 1800s when major industrialization and innovation took place first in Great Britain then throughout the World (Chen, 2019).

IR 1.0: Industrial revolution 1.0 is deliberated as the starting of industry culture, which promoted equally on quality, efficiency, and scale (Rojko, 2017). This period starts from the 18th century during which time mechanical products were started to be to manufacture, such as power and steam machine (Bahrin et al., 2016).

IR 2.0: It began from the late 1800s when it was started by the discovery of electricity that led to the low-cost productions (Zhang et al., 2018).

IR 3.0: In the 20th century, third industrial revolution began. It brought innovations in electronic industry. Many new electronic devices were invented. Comprising transistors and integrated circuits, automated machines brought advancement and automation with the use of electronic devices and IT (Bahrin et al., 2016).

IR 4.0: The recent concept, fourth industrial revolution, has been tossed in Germany since 2011. It includes cyber-physical systems, cloud storage, and internet of things. Industrial revolution 4.0 is defined as a blend of human and technology. It has reduced manufacturers' work and efforts. Even manufacturers are being substituted by robots (Rojko, 2017).

13.10.5 Educational Revolutions

Educational technologies have been observing various changes resonating with users' social experiences and learning with the web; aligning with the various version of Web, there are many advancements that are being witnessed in educational processes, as Education 1.0 to, recently, Education 4.0 (Kothari & Chatterjee, 2014).

Education 1.0: Education 1.0 is more like first generation of the web where Teacher-centered approach was being used (Harkins, 2008). Gerstein (2014) advocated that "Education 1.0 is a type of essentialist, behaviorist education based on the three Rs - receiving by listening to the teacher; responding by taking notes, studying text, and doing worksheets; and regurgitating by taking the same assessments as all other students in the cohort." Moreover, it seems technology is forbidden in classroom (Makrides, 2019).

Education 2.0: Education 2.0 is more like web 2.0. It strengthens student-to-teacher relationship and student-to-student relationship (Gerstein, 2014). Its focus is on teaching to learn rather than being more creative (Gerstein, 2014; Puncreobutr, 2016). Students are less passive in this classroom as compared with the classroom setting of education 1.0 (Keats & Schmidt, 2007). There was less use of technology in classroom like showing videos and other content-related material (Makrides, 2019).

Education 3.0: It provides diverse, cross-institutional, and cross-cultural educational opportunities where the learners themselves play a key role as creators of knowledge artifacts that are collective and where social networking and social benefits play a robust role in learning (Gerstein, 2014). Additionally, it produces self-determining learners in society because it teaches students about digital skills (Puncreobutr, 2016).

Education 4.0: Education 4.0 is a response to the needs of IR 4.0 including internet of things (IoT), Cloud Computing, Cyber security. where human and technology are associated to aid new possibilities (Hussin, 2018) and innovation (Harkins, 2008; Puncreobutr, 2016). Fisk (2017) implies that Education 4.0 is not only about learning knowledge and skills only but also about identifying the source to learn these skills and knowledge. Moreover, the aim of Education 4.0 is "to improve the digital technological competences across all levels, and to enhance the use of digital technologies for teaching and learning" (Hariharasudan & Kot, 2018).

13.10.6 Education 4.0 in Pakistan

In Pakistan, there are many drafts and policies being made to introduce ICTs in education. Pakistan's recent educational policies (Government of Pakistan, 2009, 2017) also witnessed the promotion of ICT in teaching and learning. Moreover, in 2012, the government separately drafted a National Information and communication technology (NICT, 2012) policy that has envisioned; "A country that will be globally competitive with a modern Information Technology (IT) enabled economy and a knowledge-based Information society where strong, efficient and sustainable improvements in society, economy, culture, human resources, infrastructure, legislation, regional integration and good governance are achieved through a robust IT infrastructure and effective IT applications." Furthermore, in 2018, a new policy, Digital Pakistan (Ministry of Information, Broadcasting, 2018), has been introduced by the federal government that envisages the country to turn out to be a strategic enabler for a rapid digitalization that will impact positively on the socio-economic growth and knowledge-based economy.

The knowledge-based economy means to value the knowledge, thoughts and encourage innovations, talent and technology (Talpur, 2019). Professor Dr Atta-ur-Rehman former chairperson of HEC also said that innovation, education, science, and technology are the main four foundation stones for the socio-economic development (Atta-ur-Rehman, 2018, 2019a, b). The planning commission of Pakistan in Vision 2025 (2014) has advocated the knowledge-based economy as the main pillar of progress and stated that without improving the quality education the vision of knowledge economy cannot be actualized.

Recently in one seminar related to economy the president of Pakistan Dr Arif Alvi said, "Our youth will grab the opportunities afforded by technological developments and bring the 4th industrial revolution in Pakistan" (Ary New, 2019). Moreover, the Government allocated the budget of 1.1 billion for the artificial intelligence project. From all over Pakistan, six universities are being picked up for setting nine laboratories to carry out researches (Ary New, 2019). Pakistan's government also started a campaign of digital Pakistan and hired a woman as its head, Tania Aidrus, former Google executive. Subsequently, Jazz, a cellular company, also launched a drive in 2018 "Jazz smart school" in which they used cellular communication as one of the means to educate children, and it is positively influencing the education sector. The objectives of this school are (Jazz Foundation, 2018.): improve quality of education in public sector schools in Islamabad; promote use of technology for learning and teaching; and provide access to curriculum-aligned digital content and digital lessons to inculcate better understanding of concepts.

13.11 METHODOLOGY

This chapter discusses the research methodology utilized to conduct this study. It describes the method that is used to address the research objectives and questions of this study. In addition, research paradigm, research design, instrument development process, reliability, and validity have been briefly discussed in this chapter. Figure 13.3 shows the flow of this section.

13.11.1 Research Paradigm

The approach of this study is the mixed method one in which sequential explanatory research design is utilized. In addition, pragmatism research philosophy is aligned to this study. This paradigm advocates that there is no single way to interpret and

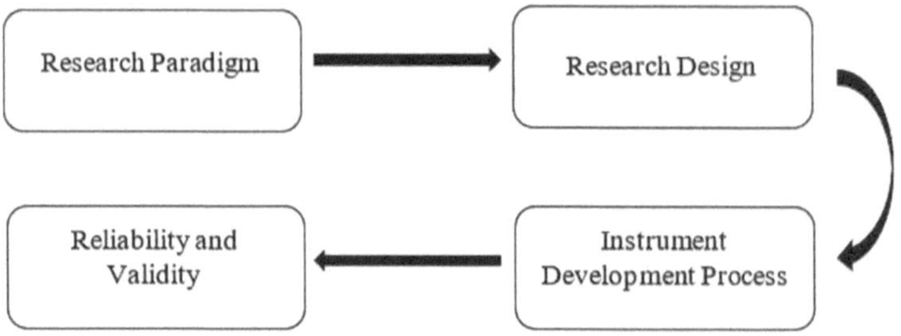

FIGURE 13.3 Flow of methodology.

explore the world and realities and undertake a research. It says that there is not any single point of view that can give the complete picture of the world and reality. There are different ways we can interpret the world and may explore multiple realities (Saunders et al., 2012).

Research question, according to the pragmatism research philosophy, plays a vital role and acts as an important factor and determinant of the research philosophy. Pragmatism can let positivism and interpretivism work together in a single research. The nature of the research question determines the combination of positivism and interpretivism. It can utilize and integrate more than one research approach, methods, and strategies within the scope of a single research (Dudovskiy, 2018).

13.11.2 Data Collection

It is a comparative analysis study in which motivation level toward Education 4.0 among Pre-Engineering and Premedical educators of community colleges and schools in rural Sindh, is assessed. As it a mixed method study, two different types of tools are used for data collection. For qualitative data collection, the survey questionnaire is used to get statistical data from 100 educators of five community colleges and three public schools. For the qualitative research design, the semi-structured interviews is conducted among eight Educators of Pre-Engineering and eight Educators of Premedical for understanding the deep knowledge about the study.

13.11.3 Data Analysis

The data collected from that questionnaire will be coded into an organizational way in Statistical Package for the Social Sciences (SPSS). Descriptive statistics (i.e., frequency analysis, measures of central tendency, and dispersion) will be used to summarize and describe the data collected from the respondents. In addition, Pearson Product Moment correlations are used to examine the relationships between the variables that were measured on the interval scale. Subsequently, statistical models (multiple linear regression) are developed to relate the dependent variable to several student demographics and factors that may influence Pre-Engineering and Premedical educators' motivation toward Education 4.0. After getting these, we can easily represent the finding in the tables, graphs, and charts form. Moreover, after transcribing the

TABLE 13.2
The Table of Specification to Determine the Motivation Level of Educators

Table of Specification for Mean

Mean Range	Level of Motivation
1.00–1.49	Highly demotivated
1.50–2.49	Demotivated
2.50–3.49	Motivated
3.50–5.00	Highly motivated

participants' interviews, the data will be analyzed using the interpretational analysis method. This will allow us to determine any themes or patterns within the interviews to explain the motivation level toward Education 4.0 among Pre-Engineering and Premedical educators of community colleges in rural Sindh. Table 13.2 shows the specification based on which we can determine the level of educators.

13.11.4 Instrument Development Process

- **Survey Instrument**

  Survey instrument is developed to collect the quantitative data. Considering the context of this study, the questionnaire (survey instrument) was adopted based on Mahdum et al.'s (2019) work with some modifications. These modifications are done with the help of available literature.
- **Semi-structured Interviews**

  Semi-structured interview guide will be developed to collect the data for qualitative part. It will be developed by considering the findings of the quantitative data. It would consist of four questions for follow-up the findings of the data collected by the survey instrument.

13.11.5 Validity and Reliability

Validity and reliability are important factors of a research to be considered before the actual data collection in the field. It is important to know that the tool you are doing to use is valid and reliable. To serve this purpose, a pilot study is a best option. But before pilot study, the tool has been given to content and construct experts to validate whether the content and the language used in the tool is valid or not.

The same is done in this research; first the tool has been given to the content and construct experts for the validation. It is given to content expert in order to validate the content of questionnaire, whether it is context appropriate and authentic content or not. After validating the content, further it is given to construct expert for validating the written language of questionnaire, whether the written language of this tool is appropriate and intends to collect required data or not. After this, a pilot study was conducted in which questionnaires were distributed among 20 participants to check the reliability of the questionnaire. In addition to this it is ensured from pilot study that the asked questions in questionnaire are well written and understandable for participants. Moreover, SPSS 22.0 software is used in order to check the reliability of questionnaire. The result of this pilot study verified the reliability of a tool.

13.11.6 Expected Outcome

The outcome of this proposed research study will help to indicate the level of motivation among educators toward the integration of Education 4.0 at Community Colleges in rural Sindh, Pakistan. It will be used as a potential reference study in the context of Pakistan for developing the framework of Education 4.0 integration in community colleges and can be further enhanced based on the potential findings of this study at the provincial level. This study will support the knowledge economy policy 2019 by the Government of Pakistan, its vision and mission.

13.12 CONCLUSION

In this chapter, we discussed that this study is based on pragmatism philosophy. Mixed method approach has been used with sequential explanatory research design according to the nature of the research question. Moreover, a survey questionnaire has been used to collect quantitative data, whereas an interview tool has been utilized to collect qualitative data. In addition, these instruments have been validated by the content and construct experts then a pilot study has been conducted. Furthermore, the data analysis tool used for quantitative data is SPSS 22.0, and for qualitative data, thematic analysis is used.

REFERENCES

Al-Hariri, M. T., & Al-Hattami, A. A. (2017). Impact of students' use of technology on their learning achievements in physiology courses at the University of Dammam. *Journal of Taibah University Medical Sciences, 12*(1), 82–85.

Ary New. (2019, March 6). Youth to bring fourth industrial revolution in Pakistan. *Ary News Network*. https://arynews.tv/en/youth-4th-industrial-revolution-pakistan-president-alvi/.

Bahrin, M. A. K., Othman, M. F., Azli, N. H. N., & Talib, M. F. (2016). Industry 4.0: A review on industrial automation and robotic. *Jurnal Teknologi, 78*(6–13), 137–143.

Bandura, A. (2010). Self-efficacy. *The Corsini Encyclopedia of Psychology*, 1–3.

Benešová, A., & Tupa, J. (2017). Requirements for education and qualification of people in Industry 4.0. *Procedia Manufacturing, 11*, 2195–2202.

Birla, N. (2018). *What skills must teachers learn to teach digital-age students?* YourStory. https://yourstory.com/2018/10/skills-must-teachers-learn-teach-digital-age-students.

Chen, J. (2019). *Industrial Revolution*. Investopedia. https://www.investopedia.com/terms/i/industrial-revolution.asp.

Christiansen, C. (2010). *Creating classroom communities of practice: Students as practioners of content (An MA dissertation submitted to the Faculty of the Evergreen State College).*

Dar, Z. (2017). *The Fourth Industrial Revolution: Implications for Pakistan*. Institute of Policy Studies.

Dudovskiy, J. (2018). *The Ultimate Guide to Writing a Dissertation in Business Studies: A step by step assistance.*

Dullas, A. R. (2018). The Development of Academic Self-Efficacy Scale for Filipino Junior High School Students. In *Frontiers in Education* (Vol. 3, p. 19). https://www.frontiersin.org/article/10.3389/feduc.2018.00019

Dunwill, E. (2016). *4 Changes That Will Shape the Classroom of the Future: Making Education Fully Technologica.* E-Learning Industry. https://elearningindustry.com/4-changes-will-shape-classroom-of-the-future-making-education-fully-technological.

Farr, F., & Murray, L. (2016). *The Routledge Handbook of Language Learning and Technology.* London: Routledge.

Fisk, P. (2017). *Education 4.0 ... the future of learning will be dramatically different, in school and throughout life.* The Genius Works. https://www.thegeniusworks.com/2017/01/future-education-young-everyone-taught-together/.

Gerstein, J. (2014). *Moving from education 1.0 through education 2.0 towards education 3.0.*

Government of Pakistan. (2009). *National Education Policy 2009.* Ministry of Education. file:///C:/Users/Iqra Kanwal/Downloads/Documents/National_Education_Policy_2009.pdf.

Government of Pakistan. (2017). *National Education Policy 2017.* Ministry of Education. file:///C:/Users/Iqra Kanwal/Downloads/Documents/National Educaiton Policy 2017.pdf.

Halili, S. H. (2019). Technological advancements in education 4.0. *The Online Journal of Distance Education and E-Learning, 7*(1), 63.

Hariharasudan, A., & Kot, S. (2018). A scoping review on digital english and education 4.0 for Industry 4.0. *Social Sciences, 7*(11), 227.

Harkins, A. M. (2008). Leapfrog principles and practices: Core components of education 3.0 and 4.0. *Futures Research Quarterly, 24*(1), 19–31.

Hussin, A. A. (2018). Education 4.0 made simple: Ideas for teaching. *International Journal of Education and Literacy Studies, 6*(3), 92–98.

Jazz Foundation. (2018.). *Jazz Smart Schools.* http://www.jazzfoundation.com.pk/our-program/education/. (2018).

Jenkins, H., Purushotma, R., Weigel, M., Clinton, K., & Robinson, A. J. (2009). *Confronting the challenges of participatory culture: Media education for the 21st century. A report for the MacArthur Foundation.* Cambridge, MA: MIT Press.

Johnson, S. (2006). *Everything Bad is Good for You: How Today's Popular Culture is Actually Making us Smarter.* New York: Penguin.

Keats, D., & Schmidt, J. P. (2007). The genesis and emergence of Education 3.0 in higher education and its potential for Africa. *First Monday, 12*(3), 3–5.

King, W. R., & He, J. (2006). A meta-analysis of the technology acceptance model. *Information & Management, 43*(6), 740–755.

Korpelainen, E. (2011). *Theories of ICT system implementation and adoption: A critical review.*

Kothari, P., & Chatterjee, A. (2014). Education 3.0 and beyond: A learner-led experience of education. *22nd International Conference on Computers in Education, Nara. Anais... Nara.*

Lase, D. (2019). Education and industrial revolution 4.0. *Jurnal Handayani PGSD Fip Unimed, 10*(1), 48–62.

Lemke, C. (2002). *enGauge 21st Century Skills: Digital Literacies for a Digital Age.*

Lineways, T. (2017). *ICT enabled education: The alchemy of mixing technology and education.*

Mahdum, M., Hadriana, H., & Safriyanti, M. (2019). Exploring teacher perceptions and motivations to ict use in learning activities in Indonesia. *Journal of Information Technology Education, 18*, 293–317.

Makrides, P. D. G. (2019). No Title. *The Evolution of Education from Education 1.0 to Education 4.0: Is It an Evolution or a Revolution? The Evolution of Education from Education 1.0 to Education 4.0: Is It an Evolution or a Revolution?*

Ministry of Information, Broadcasting, N. H. and L. H. (2018). *Digital Pakistan Policy* (p. 16). Government of Pakistan. file:///C:/Users/Iqra Kanwal/Downloads/Documents/DIGITAL_PAKISTAN_POLICY(22-05-2018).pdf.

Mossberger, K., Tolbert, C. J., & Stansbury, M. (2003). *Virtual Inequality: Beyond the Digital Divide*. Washington, D.C: Georgetown University Press.

Mourtzis, D., Vlachou, E., Dimitrakopoulos, G., & Zogopoulos, V. (2018). Cyber-physical systems and education 4.0–the teaching factory 4.0 concept. *Procedia Manufacturing, 23*, 129–134.

NICT. (2012). *National ICT Policy*. Ministry of Information and Technology Pakistan. file:///C:/Users/Iqra Kanwal/Downloads/Documents/draft-IT-Policy-revised-July-3-2012.pdf.

Nishat, R., & Michael, H. (2019). Technology in the classroom. *The News*. https://www.the-news.com.pk/print/432465-technology-in-the-classroom.

Ounis, T. (2016). Addressing the integration of ICT into teaching and Identification of the potential factors motivating teachers to use ICT. *International Journal of Humanities and Cultural Studies (IJHCS), 3*(1), 1099–1114. ISSN 2356–5926.

Player-Koro, C. (2012). Factors influencing teachers' use of ICT in education. *Education Inquiry, 3*(1), 93–108.

Puncreobutr, V. (2016). Education 4.0: New challenge of learning. *St. Theresa Journal of Humanities and Social Sciences, 2*(2), 92–97.

Rahim Sajid, A. (2013). ICTs in learning: Problems faced by Pakistan. *Journal of Research & Reflections in Education (JRRE), 7*(1), 52–64.

Rojko, A. (2017). Industry 4.0 concept: Background and overview. *International Journal of Interactive Mobile Technologies (IJIM), 11*(5), 77–90.

Saunders, M. N. K., Lewis, P., & Thornhill, A. (2012). *Research Methods for Business Students (6th ended.) Harlow*. England: Pearson Education.

Schulz, R., Isabwe, G. M., & Reichert, F. (2015). Investigating teachers motivation to use ICT tools in higher education. *2015 Internet Technologies and Applications (ITA)*, 62–67.

Sudlow, B. (2019). Review of Joseph E. Aoun (2017). Robot Proof: Higher education in the age of artificial intelligence. *Postdigital Science and Education, 1*(1), 236–239. https://doi.org/10.1007/s42438-018-0005-8.

Surendran, P. (2012). Technology acceptance model: A survey of literature. *International Journal of Business and Social Research, 2*(4), 175–178.

Talpur, M. (2019, January 10). Knowledge economy. *The Express Tribune*. https://tribune.com.pk/story/1884852/6-knowledge-economy/

Techreviews, C. (2019). *What is ICT (Information and Communication Technologies)? – Definition*.

van Laar, E., van Deursen, A. J. A. M., van Dijk, J. A. G. M., & de Haan, J. (2020). Determinants of 21st-century skills and 21st-century digital skills for workers: A systematic literature review. *SAGE Open, 10*(1). https://doi.org/10.1177/2158244019900176

Wasif, M. N., Munir, E. U., & Shad, S. A. (2012). Usage and Impact of ICT in Education Sector; A Study of Pakistan. *ArXiv Preprint ArXiv:1206.5132*.

Yusuf, M. (2011). The impact of self-efficacy, achievement motivation, and self-regulated learning strategies on students' academic achievement. *Procedia - Social and Behavioral Sciences, 15*, 2623–2626. https://doi.org/10.1016/j.sbspro.2011.04.158.

Zhang, P. (2008). Motivational affordances: Reasons for ICT design and use. *Communications of the ACM, 51*(11), 145–147.

Zhang, X., Yao, Y., Wang, Y., Xu, F., & Lu, J. (2018). Exploring metadata in bug reports for bug localization. *Proceedings - Asia-Pacific Software Engineering Conference, APSEC, 2017-Decem*, 328–337. https://doi.org/10.1109/APSEC.2017.39.

Index